Y0-AAG-476

INDUSTRIAL TRANSFORMATION AND CHALLENGE IN AUSTRALIA AND CANADA

THE CARLETON LIBRARY SERIES

A series of original works, new collections and reprints of source material relating to Canada, issued under the supervision of the Editorial Board, Carleton Library Series, Carleton University Press Inc., Ottawa, Canada.

INDUSTRIAL TRANSFORMATION AND CHALLENGE IN AUSTRALIA AND CANADA

edited by
ROGER HAYTER
and
PETER D. WILDE

CARLETON UNIVERSITY PRESS
OTTAWA, CANADA
1990

ISBN 0-88629-128-3 (paperback)
ISBN 0-88629-129-1 (casebound)

Printed and bound in Canada

Carleton Library Series 164

Canadian Cataloguing in Publication Data
Main entry under title:
Industrial transformation and challenge in Australia and Canada

(The Carleton Library ; CLS 164)
Includes bibliographical references.
ISBN 0-88629-129-1 (bound)
ISBN 0-88629-128-3 (pbk.)

1. Canada--Economic conditions--1971- .
2. Australia--Economic conditions--1945- .
3. Canada--Industries. 4. Australia--Industries.
I. Hayter, Roger, 1947- . II. Wilde, Peter D.
III. Series.

HC605.I63 1990 330.971 C90-090268-X

Distributed by Oxford University Press Canada,
70 Wynford Drive,
Don Mills, Ontario,
Canada. M3C 1J9
(416) 441-2941

Cover design: Y Graphic Design

Acknowledgement

Carleton University Press gratefully acknowledges the support extended to its publishing programme by the Canada Council and the Ontario Arts Council.

TABLE OF CONTENTS

PREFACE

This book has its origins in a workshop organized by the Industrial Geography Study Groups of the Canadian Association of Geographers and the Institute of Australian Geographers. The workshop took place at Simon Fraser University, funded by an Occasional Scholarly Conferences Grant from the Social Sciences and Humanities Research Council and by a grant from the Australian Department of Science. The Association of Canadian Studies in Australia and New Zealand also gave support. A seminar entitled "Pacific Rim, Industrialization: A Research Agenda for the 1980s," and funded by the Tri-University Pacific Rim Co-ordinating Committee of British Columbia, preceded the workshop on August 17. This seminar, which comprised papers by Allen Scott and David Angel (U.C.L.A.) on the semiconductor industry and by Mike Douglass (Hawaii) on structural changes and the Japanese economy, provided a stimulating point of departure for workshop discussions. Both these papers have been published as discussion papers by Simon Fraser University.

The purpose of the workshop, and of this book, was to stimulate a comparative analysis and discussion of the contemporary problems of industrial development and change in Australia and Canada within the context of a dynamic global economy. As semi-industrialized but rich "peripheral" countries with federal political structures established by, and for, imperial powers, notably the United Kingdom, the two countries have much in common to justify a comparative perspective. In both countries the exploitation of resources and primary manufacturers for export by capitalist-oriented societies has exerted a pervasive influence on the form and extent of industrial diversification, on spatial structure, and on fundamental attitudes and beliefs of public and private sector institutions. Market growth within the two countries has been sufficient to sustain secondary manufacturing industries not directly related to the resource sector and which have frequently been protected by tariffs, exhibit high levels of foreign ownership and remain fragmented and largely domestically oriented.

There are of course significant differences between the two countries. Nevertheless it is within the context of this legacy of resource-led export growth and dependent manufacturing that recent impacts of technological revolution, increasingly severe recessions and related crises, and multinational restructuring on actual, likely and appropriate processes of industrial transformation in Australia and Canada must be assessed. This book seeks to make a contribution to such an assessment.

We would like to end this preface by expressing our gratitude to several people. Discussions during the workshop itself were aided greatly by the perceptive comments of discussants and in this regard we are especially grateful to John Chapman (UBC), Len Evenden (SFU), Bob Galois (Vancouver), Warren Gill (SFU), Gordon Mulligan (Arizona) and John Pierce (SFU). Papers presented by representatives from the Australian Consul General, and Gary Scott (DRIE), were very much appreciated. On the social side, workshop participants thoroughly enjoyed field trips by Edward Gibson (SFU) and Warren Gill (SFU), a reception hosted by the provincial government of British Columbia and the federal government of Canada, and a dinner held at the Australian Pavilion at Expo and hosted by the Australian Consul General. We are also grateful for the important contributions to publication costs made by Simon Fraser University's Publication Committee, by the Tri-University Pacific Rim Co-ordinating Committee (Simon Fraser University) and by Carleton University Press.

We are grateful to the contributing authors for making many changes to their original manuscripts in order to both strengthen and bring up-to-date what they had to say. Finally, special thanks are due to Ida, Mary, Ben, Eric and Gerry for efforts in organizing and running the workshop and to Mary and Barb for so much dedicated typing of manuscripts.

Roger Hayter

Peter Wilde

July 31, 1990

1. INTRODUCTION

INDUSTRIAL TRANSFORMATION IN A CHANGING GLOBAL ECONOMY

Peter Wilde and Roger Hayter

Industrial transformation has been the hallmark of the past two decades at global, national and local scales (Amin and Goddard, 1986; Chapman 1987; Freeman, 1986; and Scott and Storper, 1986). Within the context of a series of economic shocks including several recessions, a period of stagflation, two energy crises and a depression, there have been major changes in the technology and organization of production, transport, communications and marketing. These changes, intimately associated with remarkably increased levels of geographic mobility of money capital, have enhanced global thinking and are contributing towards a rapidly evolving division of labour among and within regions. Moreoever, industrial transformation has provided significant challenges to the principal agents of economic decision making, namely capital, labour and the state, and their inter-relations have necessarily become more complex and dynamic.

At national levels, governments have become acutely aware of increased international competition in the markets for industrial goods, raw materials and services and of associated changes in the international division of labour. In this regard, while much of the expanded global competition and trade remains among industrialized capitalist countries, developing countries are becoming more important as a source of manufactured exports to supplement their traditional role as resource suppliers. Moreover, transpacific trade has become at least as important as trans-atlantic trade. Paralleling these geographical shifts has been a redistribution and evolution of comparative advantages among nations. In particular, as American hegemony has declined and some former "core'"regions, notably the United Kingdom, have experienced deindustrialization, West Germany and, most notably, Japan have risen to superpower status. Indeed, Japan, which within the space of four decades has shifted its export emphasis from low-wage labour-intensive products, next to capital-intensive industrial durables and then to a variety of high-wage and extremely sophisticated consumer and industrial durables *and* has become a massive exporter of equity and portfolio capital, has perhaps become the most important dynamic for global industrial change.

For Australia and Canada, two rich, resource-based "lucky" countries, industrial transformation over the past two decades poses formidable challenges. Both countries evolved in support of first British then American world hegemony, and in so doing developed close cultural and political as well as economic ties and empathies. As the relative global importance of Britain and even the US has continued to decline, and within the context of rapid economic and technological change, both Australia and Canada have been forced to reconsider their global role. Whether or not Australia and Canada should and can meet this challenge is a major theme of this volume. In particular, this volume examines and compares selected contemporary issues in industrial transformation and restructuring in Australia and Canada in the context of global dynamics and change.

There are some obvious similarities of character, history and current experience which justify a comparison of Australia and Canada. Both are huge geographically but rank low in terms of total national population. They have many parallels in their antecedents and development, especially in their former close ties to the UK. The spatial structure of their high order urban centres is remarkably similar. They each have a Westminster style of government operating in a federal system, and each displays tensions between the nation state and the local state embodied in a small number of states or provinces. Of particular importance to the thrust of this volume is the fact that their economic structures have many similarities, with exports dominated by resources and primary manufactures and a secondary manufacturing sector which grew up under tariff protection and which mainly serves the home market. Admittedly there are some marked differences in the situation, structure and institutions of Australia and Canada and these are highlighted in several of the subsequent chapters. Nevertheless, it is the broadly similar legacy of resource- based export growth and dependent manufacturing that provides the point of departure for subsequent chapters to raise questions about the direction of industrial transformation in Australia and Canada.

Contributions to this book have been divided into four sections. In the first, *Perspective*, Terry McGee begins with a discussion of the growth and character of Pacific Rim economies over the last several hundred years and of the role that Australia and Canada play within this "system". He notes the problems facing the two countries in striving to change their traditional roles and concludes, provocatively, that their best bet is simply to (continue to) hang onto the coattails of the superpowers. The next two chapters are more concerned with internal structure; Godfrey Linge outlines the evolution of the Australian political economy and emphasizes core periphery relations at national and regional scales and management/worker tensions. David Walker notes similar relations and conflicts in Canada despite what appears to be strong evidence of growing per capita income equality, at least until the 1980s, across provinces. An important question in this context is to what extent do these tensions reflect or create difficulties for industrial transformation? In the final *Perspective* chapter John Langdale notes the rapid growth of telecommunications linkages globally and how Australia's traffic has grown in tandem (including around the Asia-Pacific). An important concern is that the emerging information economy, which to an important extent is internalized within large transnational industrial and business service organizations, will exacerbate Australia's internal and external core periphery relations. Langdale believes that, in practice, outcomes are likely to be complex and that, for example, some relative cost distances may well increase. Many of the issues explored in this chapter are doubtless relevant to Canada.

The second section, *Dimensions of Employment and Social Change*, attempts, in different ways for Australia and Canada, to relate international change to regional workforce characteristics, and from this base to explore some of the social consequences of economic and workforce restructuring. In both countries unemployment has emerged as a significant and apparently long term problem during the 1980s. For the Canadian case, Glen Norcliffe and Donna Smith Featherstone document precisely the spatial and temporal variations in the growth of unemployment and the subsequent upswing between 1980 and 1984. In a novel development of the shift-share technique they verify the significance of industrial structure for unemployment and note, in particular, that unemployment is strongly related to regions with export oriented industries. Next, Jim Walmsley identifies the broad range of endogenous and exogenous changes affecting Australia, the unpredictable and uncontrollable nature of this turbulent environment, and the potential for the development of major inequalities among groups such as the aged, unemployed youth, unmarried mothers and unskilled imigrant workers. In contrast to these two chapters, which focus on the less well-off and the problems of social inequality, David Ley and Caroline Mills focus on a higher income and mobile segment of society in Vancouver. In so doing they specifically reveal the importance of the relationships between labour markets, housing markets and family status to provide a richer understanding of the dynamics of social change. In the final chapter in this section John McKay examines selected aspects of the geography of labour in Australia and notes how this geography is implicated with both the legacy of settlement history and increasing levels of international competition, especially from Asian countries.

Issues in Resource Development, are taken up in the third section with a chapter by Tom Gunton and John Richards which interprets Canadian experience in the light of the staples model of development and discusses the opportunity for resource managers to gain the rent or surplus produced. Their concern is that Canadian resources have been undervalued. In practice, Canadian governments, and indeed Australian governments as Katherine Gibson's subsequent chapter on coal development shows, have been more interested in maximizing export quantities than with maximizing rents. Moreover, in the Australian coal industry, as in many resource sectors, competition among large domestic- and foreign-owned firms, along with government efforts to promote development have consistently led to oversupply situations and problems of viability for new and existing resource communities.

Attention in the *Resource Issues* section then turns to primary manufacturing industries. Bob Fagan and David Rich's chapter on the Australian food and beverage industry relates international change, changes in corporate structures and strategies and spatial changes in employment. Their analysis of this particular industry reveals much about changes in the major resource manufacturing industries in both Australia and Canada in two main ways. First, these industries have experienced substantial technological changes which have generally increased capacity but reduced direct employment opportunities. Second, these industries have experienced major organizational restructuring, whose rationale appears as much financial as industrial, and which has resulted in an increase in the importance of corporate control, including that of conglomerates. Many of these corporations remain foreign owned. But in Australia, as has occurred in Canada, large domestically owned transnational corporations have emerged in the resource industries. In the next chapter, Richard LeHeron documents the emergence of a New Zealand forest product transnational whose recent investments in Australia and Canada, as well as elsewhere, reflect the growing interdependencies of Pacific Rim economies and which raises questions concerning the congruence of regional and corporate priorities.

The final section considers *Trade and Industrial Policy Issues.* Iain Wallace begins by assessing the Macdonald Royal Commission's inquiry into Canada's economic situation and prospects and particularly its central recommendation: that Canada seek free trade with the US (McGee's coattails policy). As Wallace observes, while the Commission's analysis lacks geographical insight and fails to consider in detail the problems of transition associated with bilateral free trade, its central recommendation has become government policy. As one alternative strategy, trade diversification is not considered practical. In fact, as Hayter notes, in macro terms Canada's global trading role outside the US remains peripheral. In the particular case of the Pacific Rim, trade has increased in relative terms but at the expense of European rather than American trade. As Peter Wilde reports, Australia has done much better in diversifying its trade partners especially with respect to the Pacific Rim. But, even more so than Canada, Australia remains a commodity supplier and, like Canada, has seemingly whole-heartedly embraced neoclassical trade principles. The costs of such an embrace are scarcely considered.

As both countries grapple with nation-wide economic problems and the wider problems of their global role, regional policy — as it has elsewhere — has been neglected and all but eliminated. Peter McLoughlin and Jim Cannon carefully document the rise and fall of regional policy in both countries and in so doing note the irony of embracing neoclassical principles as a solution to regional problems; it was, after all, market forces which led to severe regional problems in the first place. In this regard Norcliffe and Featherstone's observation of the close connection between trade and severity of business cycle problems is worth heeding.

As regional policy has declined in importance there has been a corresponding interest in technology policy in Australia and Canada. John Britton examines in detail the particular technology policy initiatives that have been developed in both countries in the context of the small firm sector. While neither country has been successful so far, in part because they have failed to recognize the "truncating" effects of foreign-owned multinationals on indigeneous technological capability, they do have options. Indeed, their ability to overcome large technology trading deficits will possibly constitute the most significant measure of their control over the industrial transformation process.

The following chapters do not offer a *comprehensive*, comparative account of contemporary industrial change in Australia (and New Zealand) and Canada; some issues are explored more in one country than in the other, and some important issues such as the nature of adjustment strategies in secondary manufacturing industries, are not examined in depth. We do feel, however, that this book identifies and analyses several major dimensions of recent industrial change and the nature of the industrial challenge facing Australia and Canada as they begin the 1990s.

PERSPECTIVE

2. THE CHANGING PACIFIC ECONOMY: CHALLENGE TO THE POST-INDUSTRIAL ECONOMY

T.G. McGee

"Such an analysis raises an important question. Has the Japanese economy 'having completed the process of' catch-up modernization gone on to become a mature (and therefore a gradually weakening) economy? We can ask the same question with regard to the developed countries of North America and Europe. Is the world economy today facing a long term decline, or will the new wave of technological innovation provide a basis for even greater growth than before?" (Kuman Shumpei, 1986, p.3).

While I cannot lay claim to expertise in the field of industrial geography, at least I have experienced in a rather oblique and protected manner the effects of the processes of industrial and economic change in Australia, Canada and New Zealand. This chapter offers what might be called, in the jargon of the day, "the contextual setting", for these processes but what I would prefer to call in my own "peasant" idiom a background paper to the more specific issues raised in subsequent chapters.

The central issue of this chapter, and in some ways of the entire volume, is intimated by the above-mentioned quote from Shumpei (1986). Simply put, it is as follows: will Australia, Canada and New Zealand be able to *position* themselves, or *be positioned* (in the sense that many of the processes positioning them are beyond their control), so as to take maximum economic advantage of the current phase of economic readjustment in the world economy as it is reflected in the Pacific region? In order to assess this issue, the chapter is divided into four parts: what is meant by the Pacific region and the idea of a Pacific economic system; the historical patterns underlying the emergence of the Pacific economic system; the effects of the Pacific economic system on contemporary economic and social processes in the region; and the implications for Australia, Canada and New Zealand.

2.1 DEFINING THE PACIFIC REGION: A CURIOUS TASK

There have been innumerable attempts to define the Pacific region. It is a daunting and curious task to define regional components in an age of the global economy, especially bearing in mind the immense geographical and cultural diversity of Pacific Rim countries (see McGee, 1983). The region conjures up a kaleidoscope of images:

— The word "Pacific" connotes the relaxing image of coral atolls,white beaches waving coconut trees and suppliant Pacific people.

— The beauty image of the Javanese countryside with rice-plants-black—in the mirror lakes of rice fields that climb up the terraced sides of mountains.

— The progress image of downtown Singapore, or Hong Kong with office towers—temples of commerce multiply to immitate the Wall Street of New York.

— The poverty image of refugees clustered in overcrowded camps with potbellied children peeking from behind the sarongs of adults.

— The political image of Red Guards marching under flowing red banners—choreographed iconography so beloved of the political image makers.

— The festival image of Chinese dragon dancers on Pender Street on Chinese New Year. The streets are full of Chinese faces. Is this Hong Kong or Vancouver?

— The philosophical image of a Japanese temple garden in Kyoto—stone rakes in wave-like lines—an ocean of pebbles. Sit still and watch—eventually the pebbles begin to move like waves striking a beach.

One could go on but let me make my first point. Most of us perceive this vast region through such images. These diverse images are further reinforced by immense geographical diversity and heterogeneity. Within the region lies the world's largest country, China, with a population of more than one billion. The region also has some of the world's smallest countries such as Nauru, which has less than 12,000 people. Within the region are some of the most ancient nations such as Korea, and some of the most recent such as Vanuatua, formerly the condominium of New Hebrides, which received independence in 1980. Within the region there is immense linguistic and cultural diversity ranging from the Catholicism of the Latin American countries to the Islam of Malaysia and Indonesia. In addition, there are at least 50 political units comprising ideologies ranging from the socialist regimes of the USSR and China to neo-conservative regimes as Chile.

Notwithstanding its diversity and geographical extent, there are those who see great unity in the region. In 1900, the American Secretary of State John Hay, wrote that the "Mediterranean is the ocean of the past, the Atlantic is the ocean of the present, and, the Pacific is the ocean of the future". Today, those who feel this prophecy has come true point to the fact that already, interregional trade throughout the Pacific exceeds that among the Atlantic nations and that many of the more dynamic Asian countries have experienced over the past several decades the fastest national rates of growth in the world. The combined gross national product (GNP) of the East Asian countries alone will exceed that of the European Economic Community (EEC) by 2000 A.D. Some Japanese economists even envisage a Pacific Basin linked together in a manner not unlike the EEC. This perception of Pacific region unity is based upon increasing economic growth, trade and interdependence. Whether growing economic links will lead to political and cultural ties remains to be seen.

2.2 HISTORICAL PATTERNS OF INCORPORATION OF THE PACIFIC REGION INTO THE WORLD SYSTEM

What other unifying elements characterize this vast region? Historically, the region shares the common experience of a periphery, (a periphery rich in tradition, culture and art), and of incorporation into a global system of economic and political relations which historically has been dominated by capitalism emanating from Europe. This process has been temporarily and regionally uneven, assuming different forms in different parts of the Pacific. Four phases of incorporation into the global system can be delineated.

The first phase is one of *plunder*. It begins with the conquistadors' plundering of the treasures of the Americas and continues with the spice trade of Asia, and the sandalwood and whaling trade of the Pacific.

A second phase is of *colonization* in which the European powers attempt to establish political control over their regions of plunder. The Pacific is carved up between the European powers who introduce their own institutions, religions, language, and culture. There are three types of society that emerge as the result of these impacts: the white settler economies including Australia and Canada; the genuine colonies directly under foreign rule of which most Latin American and many Asian countries are typical; and the countries that avoided direct colonial rule like Japan, China and Thailand.

The third phase is one of *nationalism* in which political control passes to national units and new policies of internal development are put in place. This phase, of course, is still of major importance and is characterized by the major role played by the state in economic development. Finally, there is a fourth phase of *economic incorporation* involving substantial economic restructuring. This phase is characterized by increasing interdependence, specialization and the growth of the control of transnational corporations based in Japan, the US and Europe within the region.

Stilwell (1980) has suggested that there are five interrelated processes bringing about this restructuring. First the world economic system is increasingly dominated by transnational corporations. Since 1950 transnationals have grown rapidly in advanced capitalist societies and in the Third World (see Fagan, McKay and Linge, 1981). As Langdale explains in Chapter five there has been a rapid internationalization of service activities, greatly aided by the development of computer-based electronic information systems, in such fields as banking, finance, law, advertising and various corporate- related services (see also Cohen, 1981). Firms in these activities operate in a system of international financial markets less subject to the control of the central banking system (Clairmonte and Cavanagh, 1984; Daly and Logan, 1986; Daly, 1987; and Thrift, 1986). A system is evolving in which large multinational firms have much greater flexibility through a wide variety of measures to move their operations throughout the world; moves which are actively encouraged by most nation states of the region.

A second process which has greatly facilitated these trends has been the technological revolution in communications, personal travel and transport, including the growth of bulk carriers, containerization and wide-bodied jets which have facilitated the mobility of financial capital, reduced the "pull" of resources, facilitated information exchange and, in general collapsed time-space relationships, making place less important.

A third process outlined by Stilwell is what he labels the "relative disintegration of national economies". The increasing internationalization of the world economy means that particular regions of the national economy have become more closely integrated with the international system. This poses increasing tension for the nation state, as the conflicts between western Australia and Queensland, British Columbia and Alberta and the federal governments of Australia and Canada show only too well.

These processes are related to a broader process sometimes defined to include all of the above and is referred to as the New International Division of Labour (NIDL). In its more precise form the term has been used to describe the process whereby industries with high labour content such as textiles, shoe-making, clothing and electronics have been relocated from the advanced countries to certain Third World countries where cheap labour abounds and where an institutional environment provides a high degree of security and encouragement for investment. The institutional forms which this decentralization of manufacturing takes are quite diverse, ranging from the often described export zones to subcontracting and even complementation schemes where part of the production cycle is relocated in Third World countries (see Frobel, Heinrichs and Kreye, 1980).

During the 1960s and 1970s there was a considerable growth of export-oriented industries in the Third World. In the period between 1965 and 1975 these exports increased from US$4.6 billion to US$33.2 billion at the rate of 16.3 percent compared to 10.8 percent in the developed countries. Of course, this increase was geographically unevenly distributed with over 50 percent of the increase occurring in the major export platform economies of Korea, Taiwan, Hong Kong and Singapore. Between 1963 and 1976, for example, the percentage of world export manufacturers contributed by these four countries rose from 1.37 to 4.10 percent. In the period between 1970 and 1979 this concentration was much higher for certain products. Hong Kong and Korea alone accounted for almost 70 percent of the increase in exports of clothing to developed countries. For miscellaneous light industrial products, especially electronics, the figure was almost as high. Finally, it should be noted that while the product mix of these exports has been dominated by textile, clothing and electronics, there has also been a considerable increase in machinery, transport and chemical exports.

A fourth process, which is partly related to technology but much more clearly related to the fluctuating rates of exchange and monetarist policies of states in the region, involves the considerable flows of investment both direct and indirect which have accelerated in the region since 1970. In recent years, this trend has been dominated by the massive movement of Japanese capital into the securities markets of the US and Canada (see Ninomi, 1985).

A final process is the increased mobility of populations within the region which is reflected in increased tourism, refugee movements, contract labour employment and immigration.

2.3 THE EFFECTS OF THE INCORPORATION PROCESS OF THE PACIFIC SYSTEM

The end effect of these processes is to position the countries of the Pacific region within a global system in which their national economies are to a considerable degree shaped by their relative position in this system. The role of both national and multinational corporations and of governments are all crucial in this positioning. Whether the aggregate effects lead to the emergence of a Pacific economic system in which it is more beneficial to an individual state to increase its economic linkages within, as opposed to outside, the region is very difficult to establish.

There are considerable problems in measuring the emergence of this Pacific Rim system. Apart from the problem of defining the spatial components of the system, there is the question of measuring the degree of interaction within the system. For instance, the use of economic statistics such as value of trade is subject to many qualifications. Similarly, the geopolitical realities of the Pacific Basin cut across emerging interdependencies. However, it can be argued that there are five components of a Pacific economic system: the supereconomies of Japan and the US; the white settler countries of Canada, Australia, New Zealand; the Newly Industrializing Countries (NICS); other Asian countries; and the large Latin American countries.

There are four main features of the system (Tables 2.1 to 2.4). First, the supereconomies of Japan and the US are dominant in the system. In the period since 1965 they have been responsible for producing three quarters of the Gross Domestic Product (GDP), 76 percent of the value of industrial output and over 80 percent of value of manufacturing output. Second, the system has been remarkably stable over the last 20 years. Despite the much heralded rise of the NICs, the largest shifts in GDP and value of industrial and manufacturing output have been in the considerable increase in Japan's share of the Pacific Rim output at the expense of the United States. Essentially, the system has witnessed a relative shift in importance within the supereconomies, which has made very little difference to their total share of the Pacific Rim's output. These developments have not been without growing friction between the supereconomies as the "trade war" of 1986 attests. But the Pacific Rim economic system still is dominated by these two economies.

TABLE 2.1

PERCENTAGE GROSS DOMESTIC PRODUCT AND POPULATION OF PACIFIC RIM COUNTRIES: 1965-1983

	1965 GDP (pop.) Percent		1979 GDP (pop.) Percent		1983 GDP (pop). Percent	
1. Supereconomies						
USA	68.7	(14.3)	52.6	(12.3)	56.8	(12.2)
Japan	9.1	(7.2)	21.8	(6.3)	18.4	(6.2)
	77.8	(21.5)	74.4	18.6)	5.2	18.4)
2. White Settler						
Canada	5.2	(1.4)	5.1	(1.2)	5.6	(1.2)
Australia	2.3	(0.8)	2.9	(07)	2.9	(0.8)
New Zealand	0.6	(0.1)	0.4	(0.1)	0.4	(0.1)
	8.1	(2.3)	8.4	(2.0)	8.9	2.1)
3. NICs[1]	0.9	(3.3)	2.6	3.4)	2.8	(3.4)
4. Other Asian[2]	8.2	(64.8)	8.5	(66.0)	.0 (	67.7)
5. Latin American[3]	5.1	(7.6)	6.2	(8.5)	5.9	(7.9)
TOTAL	100.1	(99.5)	99.5	1(00.2)	99.5	100.2)

[1] NICs: Korea, Hong Kong, Taiwan, Singapore

[2] Other Asian countries: People's Republic of China, ASEAN (except Singapore)

[3] Latin American: Mexico, Columbia, Peru, Chile, Argentina

Source: World Bank

TABLE 2.2

PERCENTAGE DISTRIBUTION OF VALUE OF INDUSTRIAL OUTPUT CONTRIBUTION TO PACIFIC RIM COUNTRIES GDP

	1965	1979	1983
1. Supereconomies			
USA	69.5	50.0	53.5
Japan	10.4	25.6	22.8
	79.9	75.6	76.3
2. White Settler			
Canada	4.7	4.7	4.8
Australia	2.5	2.4	2.2
New Zealand	0.2	0.2	0.1
	7.6	7.3	7.1
3. NICs	0.7	2.5	3.2
4. Other Asian	7.5	8.1	8.6
5. Latin American	.7	6.9	5.6
TOTAL	100.4	100.4	100.8

Source: World Bank

TABLE 2.3

PERCENTAGE DISTRIBUTION OF VALUE OF MANUFACTURING OUTPUT IN PACIFIC RIM COUNTRIES

	1965	1979	1983
1. Supereconomies			
U.S.A	75.7	54.2	57.0
Japan	11.0	28.1	26.0
	86.7	82.3	83.0
2. White Settler			
Canada	4.5	4.1	4.3
Australia	2.5	-	1.8
New Zealand	-	O.4	0.5
	7.0	4.5	6.6
3. NICs	0.6	3.1	3.7
4. Other Asian	0.9	1.9	2.7
5. Latin American	4.7	8.1	5.0
TOTAL	99.9	99.9	101.0

Source: World Bank

TABLE 2.4

PERCENTAGE SHARE OF INDUSTRIAL LABOUR FORCE IN PACIFIC RIM COUNTRIES

	1965	1979	1983
1. Supereconomies			
USA	43.2	35.1	4.5
Japan	21.9	22.2	1.7
	65.1	57.3	56.2
2. White Settler			
Canada	3.9	3.4	3.4
Australia	2.8	2.3	2.3
New Zealand	0.6	0.5	0.5
	7.3	6.2	6.2
3. NICs	4.3	9.0	8.9
4. Other Asian [1]	9.4	12.8	13.3
5. Latin American	13.7	4.8	15.2
TOTAL	100.0	100.1	99.8

[1] Excludes China

Source: World Bank

Third, despite the very considerable focus of attention on the economic growth of the NICs as major components of their own "economic transformation", they still represent a comparatively small proportion of the total Pacific rim output. In this respect it is interesting to note that the "white settler colonies", despite their real concerns about their industrial and economic performance, have remained stable contributors to the system.

Fourth, while the unemployment issue has been central to the so-called Asian challenge for the "supereconomies and white settler colonies," Table 2.4 shows that it is the US which has carried the major part of employment loss. In this regard, the loss of employment in such industries as textiles in the US, Canada and elsewhere is particularly noteworthy and has been the cause of much of the tension. It is also true that over the last few years the increase in informal employment in such industries in the US has not been included in official statistics (see Sassen-Koob, 1984).

Of course these tables are oversimplified and need to be filled out with data on, for example, investment, commodity flows and tourism. They begin to delineate, however, the skeleton of the Pacific Rim system, a system dominated by the two supereconomies of the US and Japan.

2.4 THE EFFECTS OF THE PACIFIC RIM SYSTEM ON THE POPULATIONS OF THE REGION

The chapter now considers how this incorporation of the Pacific periphery into the global system of economic and other relations has affected the human condition of the peoples of the Pacific.

First, it has had a destructive effect on the aboriginal peoples in the Pacific Rim. I use aboriginal in the sense of original or first inhabitants. From the Inuit in the North of Canada, to the Indians of America, to the Maoris of New Zealand, the Aborigines of Australia, the Indians of Latin America, the Semoi of Malaya, and the Dayak of Borneo—all have experienced a destructive impact in which their land, their resources, and their way of life have been destroyed. Only in the last 20 years have various coalitions of native people begun to join together to make their rightful claims to their heritage. This is one common element of the human condition of the Pacific Basin.

Second, the incorporation process has been a factor in the introduction of various forms of medical improvements which are responsible, in part, for a rapid surge of population in the region—which is the underlying time bomb in the relationship between population increase and food and energy supply. Even the most optimistic forecasts perceive the region doubling its population by 2100, and this assumes the widespread acceptance of the one-child family in China, the giant of the region, and increasing acceptance of birth control throughout the region. Much has been made of the "green revolution" in agriculture as the solution to the food supply problem, particularly in Asian countries. But while it has largely fulfilled its promise of food self-sufficiency in basic cereals, it has proved a costly task, and the environmental side effects are still being assessed. Moreover, there have been significant changes in patterns of urban food consumption which have encouraged the importation of basic staples such as wheat from Western countries notably, Canada, the US and Australia (see MacCleod et al., 1987). Unfortunately, the rate of importation has not been at the desired levels and the resulting food problems have fallen most heavily on the urban poor and rural poor without land. The 'green revolution' technology is, of course, reliant upon "high energy" inputs such as chemical fertilizers and pesticides. And these inputs involve taking the control of the production process increasingly out of the hands of the farmer and putting it into the hands of the experts and of the chemical companies such as Imperial Chemical Industries. It is a fragile and dangerous situation which affects the human condition of most inhabitants in the Pacific Rim.

Third, the incorporation process has accelerated the patterns of urbanization in the region. While the majority (75 percent) of the people in the developing areas still live in rural areas, the cities of the region are growing rapidly. In this regard, the larger cities in particular are linked by international travel and have become the locus for transnational and national capital. Other technologies such as communication technology emphasize their growing importance. Today the region has the "dubious" reputation of including three of the larger cities of the world: Mexico City (14 million), Tokyo (12 million) and Shanghai (12 million). But other cities, such as Singapore, Hong Kong, Los Angeles, Honolulu, Jakarta, etc. act as the facilitating framework for internationally mobile capital and investment. A new class of transnational technocrats move about within and between these cities, managing the new world economic order. Evidence of their thinking is found in the Club of Rome, the Club de Geneve and the Trilateral Commission. There is increasing convergence in the built environment and class structures of these nations. You may stand at Pender Street in Hong Kong, in Raffles Square in Singapore, in the banking district of Tokyo or in downtown Los Angeles and see the same skyscrapers. You can walk in the department stores and see the same goods for sale. There is thus an increasing convergence of life styles especially among the middle and upper classes.

On the other hand large sectors of urban populations still live in substandard housing, squatter settlements and engage in low-income occupations. This social structure is reproduced as one moves down the urban scale to the small towns. Recent research emanating from Friedmann and Wolff (1982), Cohen (1981) and other theoretical works of the 1980s (Armstrong and McGee, 1985; McGee, 1986; Rimmer, 1985) argues that the "global cities" are major "staging bases" in the spatial reorganization and articulation of markets of the global economy. The linkages between these global cities — in terms of flows of money, commodities and information — are often much more important to the viability of the city than the linkages to the nation of which it is part. From this perspective, Friedmann and Wolff (1982) identify Los Angeles, Tokyo, San Francisco, Singapore, Hong Kong, Taipei, Manila, Bangkok, Seoul and Mexico City as being major components of the system of global cities.

Fourth, this latest phase of the world economic system is bringing with it a new international division of labour. The restructuring of industry in the developed countries of the world core has led to the relocation of labour-intensive assembly operations in offshore locations in countries such as Korea, Taiwan, Hong Kong, Singapore, Mexico and Malaysia. Often, such activities are encouraged by the developing Pacific countries eager to create labour opportunities. Up to 70 percent of the workers are women, and clothing and electronics prevail. Wages are up to 10 times lower than in developed countries and women workers often receive only two thirds of the wage rates of male workers. Even China is encouraging this trend by setting up special export zones, for example, Shenzhen next to Hong Kong. This process is also occurring in agribusiness where the great international food companies such as Del Monte are extending their operations from Latin America into parts of Asia such as the Philippines and Thailand. The conditions of agricultural workers in these enterprises is often much worse than those of the urban proletariat (Scott and Storper, 1986).

A fifth aspect of these developments in the Pacific region is the increasing mobility of labour throughout the region. Canada receives large numbers of immigrants from Hong Kong and Asia. Mexican immigrants, legal and illegal, flood into the US and Philippine workers move to the Middle East. Another aspect of the international movement is tourism. Over 46,000 tourists a day flow into Hong Kong spending up to US$2,000 a visit; most are looking for food and shopping. In other cities the motivations of tourists differ. In Bangkok, a city of four million, over 200,000 prostitutes are available for special "exotic" tours from West Germany and Japan.

2.5 CONCLUSION: CAN THE WHITE SETTLER COUNTRIES CATCH THE THIRD WAVE?

To conclude I briefly raise some issues specifically concerning Canada, Australia and New Zealand which I have linked together under the rubric of the "white settler countries". As a point of departure it is useful to refer to the core-periphery model of the world system put forward by Frank, Wallerstein and others. A problem with this model is how to incorporate so-called semi-peripheral countries such as Canada, Australia, and New Zealand. Essentially, the model contrasts a core of capitalist states made up of highly industrialized states and a periphery consisting of dual economies in which the forces of production have not fully developed. The major relation between these two components of the system comprises the highly unequal transfer of value from the periphery to the core. In the earlier versions of this model that relationship leads to persistent underdevelopment and inequality.

Countries such as Canada and Australia have developed within this system and their existence has presented some problems to the theoretical basis of the model. These problems have been further compounded by the rapid economic growth of the NICs. To writers such as Logan (n.d.) and Fagan et al. (1981) the existence of these special cases can be explained through the particular environment, resource base and form of capitalist investment that have historically characterized their growth. But these conditions are insufficient to invalidate the general theory. While I agree with these diagnoses of the historical reasons for the emergence of these semi-peripheral countries I am concerned that these attempts reflect rather naive "spatial" thinking and an excessive concern with locating countries in some continuum between the most peripheral and the most "core."

This sort of exercise really takes us away from what should be the central concern which is to analyse the manner in which countries are incorporated into an international economy. The core-periphery theory really deals rather inadequately with the kind of international processes of capitalist penetration (led in many cases by transnational firms) that are creating an international and national class bound together by shared consumption patterns and commonalities of class interest that transcend core-periphery divisions. Looked at simply, it may be more important to look at McDonalds as an international marketing franchise located in Australia than as an economic activity which will take some different form in a semi-peripheral social formation than a core one. A theory of capitalist incorporation which accepts diversity in these patterns does not ignore the unequal relationships of countries within the capitalist system but it is much more comprehensive than the core-periphery theory.

This discussion is particularly relevant to countries such as Canada and Australia, for it emphasizes the manner of incorporation into the international capitalist system as primary resource suppliers in which their efforts to become industrial nations by policies of protection and import substitution were successful at one particular phase of the international capitalist system. At the present time, however, there are real questions as to whether Australia and Canada can maintain, let alone expand, their manufacturing bases. The implications of these developments suggest two possible scenarios over the next 20 years.

The first is the *post-industrial* scenario (sometimes called the Third Wave). The Bell model of post-industrial society has increasingly come under attack because of its lack of recognition of the "dark underbelly" of the post-industrial society—that is the emergence of "downgraded" poorly paid, often unemployed, often immigrant groups in the population. Allowing for this limitation, the post-industrial scenario assumes a resurgence of the world economy, with radical restructuring involving an accentuation of the trends of the last 15 years in the US in which employment in manufacturing has stagnated relative to the service sector. This scenario suggests an increase in incomes and leisure activities.

The features of this process have been much described and debated in the US and I will not review them here. The process involves both sectoral shifts in production, spatial shifts in investment and considerable change in labour markets. However, in the US Noyelle and Stanbeck (1983) and Sassen-Koob (1984) have emphasized the significance of the growth of producer services (provided along with goods)—advertising, research information—which are necessary for the growing product differentiation that has characterized consumer markets (Bluestone and Harrison, 1986; Piore and Sabel, 1986; Scott, 1986). This trend is occurring in large urban centres and some regions while others become more deindustrialized (Clark, 1986). At the same time the middle-class blue-collar workers are being squeezed for jobs and income, and lower level-services and labour-intensive industries are being downgraded into non-union, low-paid occupations. This is what Scott (1985) calls the "disintegration of industry" including breakdowns in the size of production units.

The dignity and security of the worker, the conscience of the union movement, are becoming increasingly more difficult to obtain. This is creating dualistic labour markets which have some similarities to the classic models of Japan. Japan, it seems clear, will move rapidly along these lines, in fact, reinforcing a system that has historically characterized its industrial growth (Douglass, 1986 and 1987; Wonoroff, 1985; Yutaka, 1986). Over the the last 20 years it has been creating the infrastructure to make this shift (as yet with limited employment restructuring) but it is now moving into a new phase of internationalization in which more and more basic goods manufacture will be located off-shore in a diverse range of locations in Taiwan, Korea, US, Europe, Canada and Australia. The Ministry of International Trade and Industry suggests this would be a loss of one million jobs in Japan over the next 10 years which would be about five per cent of the labour force in industry today.

The second scenario rests upon the assumption of a deepening world *recession*, which some theorists of cycles of accumulation and decline in the world economy feel fits what has been going on since the mid- 70s. In such a situation one can imagine increasing protectionism, fuelled by the state's need to contain the welfare crises; "stagnating industry" and increasing regional disintegration including more policy "ad hocism."

The implications of these developments for the role of the white settler countries are most interesting to think about in terms of the first scenario. Hence, the rather crucial question, "Can the white settler countries catch the Third Wave?" As I view it, Australia, Canada and New Zealand are now attempting to position themselves so as to take advantage of this "Third Wave". Thus Sydney and Vancouver are both attempting to develop service functions to offer competitive positioning among the global cities of the Pacific economy.

The success of these endeavours will primarily rest upon how well the "white settler" countries can climb onto the coattails of the supereconomies. Canada may have already recognized this fact by entering into free trade with the US. Of course, this ability to climb onto coattails poses challenges to national identity and the mind set of the decision makers of the white settler countries. Let me be provocative and conclude with the observation that given the current dominance of the supereconomies in the Pacific economic system Canada and Australia have few choices but the coattails approach.

3. THE AUSTRALIAN ECONOMY IN TRANSITION

G.J.R Linge

This chapter is divided into four substantive sections. The first briefly describes the development of the political economy of Australia, including the all-important division of responsibilities between the state and the federal governments. The second broadly covers the events which have shaped the Australian economy since the 1930s and have led to a new focus on countries around the Pacific Rim, especially in southeast Asia. It is noted how this, in turn, has drawn attention to two related dilemmas. One is the changing trade-off between the high levels of protection enjoyed by Australian manufacturing (by tariffs, quotas and other means) whilst finding ways of developing export markets in countries which strongly object to such impediments to their exports. The other is the traditional perception of Australia's role as largely an exporter of minimally processed farm and mineral products rather than as a developed country exporting sophisticated goods and services. The third section outlines some of the main changes that have taken place in the spatial organization of the Australian economy, while the fourth considers the stresses appearing in the social fabric.

3.1 DEVELOPMENT OF AUSTRALIA'S POLITICAL ECONOMY

The period from the arrival of the first European settlers in 1788 to Federation in 1901 saw the delineation of the fundamental attributes of Australia's economic and human geography and the development of a parochial rather than a national attitude which is perpetuated in many ways. For example, in the colonial era tariffs were imposed by each colony on most incoming products whether they originated elsewhere in Australia or overseas. Roads and railways (mostly government-built, owned and operated) fanned out from each mainland capital to channel imports and exports there, rather than to the ports of neighbouring colonies. Only a handful of instances has been documented where firms operated in more than one colony and, in general, business identified itself with a particular part of the continent rather than with Australia as a whole (Linge, 1979). Then the colonies entered Federation in 1901 retaining as many of their rights as they could, by curtailing federal powers, and also by their competitive and assertive attitudes at the level of the individual state. In short, even at the end of the nineteenth century each of the colonies had closer economic ties with the United Kingdom—despite being to all intents and purposes politically independent of it on a day-to-day basis—than with any other part of Australia. Britain remained the main market for the bulk of each colony's exports—wool, meat, leather, copper and gold—and the main source of its manufactured goods and equipment.

Within each of the six colonies the capital city clearly dominated political and economic affairs and this domination was reinforced over time. The increasing scale of business organization and technological efficiency encouraged the concentration of manufacturing and services in the capital, often entailing closure of operations elsewhere in the colony. In addition, governments contrived to increase the capital's dominance by, for example, manipulating railway freight rates. These trends invigorated the antagonisms and sustained the schisms between people living in each metropolitan centre and the rest of each colony and between farmers and manufacturers. This is a second dimension of inherited attitudes and tensions, which is crucial to an understanding of the present-day Australian political economy.

One vital outcome of these tensions was that the federal government was left under the Constitution with overall powers relating to external matters (like foreign affairs, defence, tariffs, and immigration) and internal responsibilities that no one else wanted (like running the post and telegraph services). The most immediate impact of Federation was to eliminate internal tariffs. Indeed, the Constitution (Section 92) specifically stated that:

> On the imposition of uniform duties of customs, trade, commerce, and intercourse among the states, whether by means of internal carriage or ocean navigation, shall be absolutely free.

This Section has been legally tested many times. In theory it has been upheld; in practice it is weakened by state legislation that imposes, for instance, seemingly arbitrary controls on the labelling, packaging, content and other attributes of products which are claimed to be necessary for, say, health and safety, but which "incidentally" help local enterprises to maintain their market share. The competitive and assertive nature of the former colonies continues to have a powerful effect on many aspects of Australia's political, economic and social life. Differences in regulations between states affect everything from the additives allowed in a can of peas to the permissible axle loadings for trucks, from the curricula in primary and secondary schools to road rules, and from the processes of wage negotiation and arbitration to the adoption of daylight- saving time. In effect, a country with a total internal market of only 16.5 million people is fragmented into eight rather different sub-markets (the six states, the Northern Territory and the Australian Capital Territory). Thus manufacturers have to decide whether to compete in only one sub-market or to make products that meet all the different sets of specifications.

The states retained (and have continued to jealously protect) complete authority over resources like land, minerals, forests and water and have used this power to bargain among themselves for industrial developments. To a greater or lesser extent all states over many years have made resources available to individual producers on the formal condition or informal understanding that downstream processing plants would eventually be established in that state (Linge, 1967). Moreover, until 1986 each state awarded contracts to local producers even if their tenders were as much as 10 percent higher than those from elsewhere, thus creating yet another disincentive to the achievement of national economies of scale. The concern by each state to acquire a full "portfolio" of industrial and commercial activities has had two important consequences. First, along with high tariff protection, it has created gross inefficiencies because of the duplication of facilities operating much below international levels of annual output. This has cut across many of the initiatives that federal governments have tried to promote from an Australia-wide perspective. Second, it has strengthened the hand of individual firms at the negotiating stage since they have been able to play-off the menu of assistance measures offered by one state against that of another, thus further fostering irrational investments at a national scale and increasing long-term public expenditure commitments in services and infrastructure.

The states retained the right to raise revenue through taxes on personal and company income, as well as from miscellaneous charges. In 1942, as a wartime measure which has never been repealed, the federal government took over responsibility for raising revenue from personal and corporate income. The government pays part of this, plus income derived from other sources such as tariffs, to the states in the form of general-purpose and specific-purpose grants. Approximately one third of federal revenue is now allocated in these ways, which makes up about half the total income of the states. In the past there was much special pleading by the states collectively and individually about the level of funding, but in recent years the federal government has adopted a much firmer stance. Thus, in the light of Australia's economic problems (discussed later) the federal government in May 1987 cut state funding (grants and global borrowing limits) for the fiscal year beginning on July 1, 1987 by 7.5 percent in real terms. As part of this more stringent attitude it reduced the level of allowable borrowings by the states and their host of statutory authorities by 21 percent, since these have become an important component of a fast-growing public sector indebtedness, and an important factor in Australia's mounting net external debt. This firm control of state borrowing has continued.

The 90 years since Federation, however, have seen a series of attempts by successive federal governments to bring their influence to bear on many issues. But even here the states' rights are constantly argued, sometimes complicated by loyalty to a (national) political party, and outcomes tend to be less clear than federal governments would wish. For example, minimum rates of pay for most workers are prescribed under federal awards but some workers are also covered by various state industrial tribunals. Similarly, while the federal government has necessarily become involved in many aspects of the financing and control of activities which are Australia-wide in their implications (especially in such fields as health, social services, welfare, education and air safety), there is much room for disputation and political point-scoring as to which level of government is to blame for inadequacies or can claim the credit for initiatives. The states are very sensitive about upholding their real or perceived rights under the Constitution, with any threatened federal intrusion being seen as the thin end of the wedge.

Many examples of these tensions between federal and state governments could be listed although, over the years, the nature of the issues in dispute has changed. Currently, one of the most important is the lack of wholehearted co-operation between state and federal investigatory and enforcement authorities in relation to organized crime, ranging from drug rings to company tax avoidance and illegal immigration. Some of the most intractable and acrimonious confrontations involve environmental issues. For instance, in the 1970s and 1980s the federal government used its powers over what can be legally exported from Australia as a means of stopping particular mining and forestry operations, the products of which were were almost entirely for use overseas. The state governments, anxious for economic development, see this kind of intrusion as an improper—even blatant—extension of federal powers; others see it as a necessary and desperate last-ditch attempt to save environmentally fragile areas. The chances are that such disputes will increase because only the federal government has the power to sign international treaties and conventions but these place on it the onus to ensure that such agreements are observed throughout Australia.

3.2 ECONOMIC AND POLITICAL DILEMMAS

Australian trade policies during the 1930s adhered to the 1932 Ottawa Agreement and to the preferential principles operating in the British Commonwealth. In 1939 bulk purchase arrangements (fostered by the probable outbreak of hostilities in Europe) seemed likely to tie Australian trade even more closely to the United Kingdom, a prospect that caused resentment in the United States which believed that world trade should be freer. In 1942 both the United Kingdom and Australia signed Article VII of the Mutual Aid Agreement with the United States: this ensured immediate aid under the lend-lease system in exchange for a commitment to review the Commonwealth preferential system after the Second World War and the review helped to pave the way for the General Agreement on Tariffs and Trade (GATT) in 1948.

The war also revived Australia's concern about its vulnerability to armed invasions from Asia and the illegal occupation of the very sparsely settled north. With a population of only seven million and a falling net reproduction rate, action was taken before the war had ended to revive the flow of migrants from Europe and, in particular, from the United Kingdom. Between 1947 and 1971 there was a net gain from immigration of 2,300,000 of whom 42 percent came from the United Kingdom alone. These settlers and their children accounted for 59 percent of total population growth during the 1947-71 period. Wartime experiences had hardened attitudes against Asians (particularly the Japanese) and the emphasis was firmly on settlers of European and Anglo-Saxon origin: of the 12,756,000 people in the country in 1971 only 1.3 percent had been born in Asia.

During the 1950s and 1960s Australia was faced with a growing dilemma. On the one hand, old and new social and cultural links with Europe were being maintained and developed as the stream of migrants continued while, at the same time, there remained a mental blockage to most things

Asian. On the other, some of the economic and trade links with Europe were weakening as intra-European ties strengthened and greater self-sufficiency in agricultural products was encouraged. The negotiations by the United Kingdom in the early 1960s to join the EEC was a clear warning of the changes to come.

Some influential members of the (Conservative) federal governments of the period recognized the problems that were looming but found it hard to persuade their colleagues, the trade unions or employer groups about actions that should be taken. Nevertheless, since the share of Australian exports going to Europe was declining (from 58.9 percent in 1950-51 to 22.7 percent in 1970-71) alternative markets had to be found and the Asia-Pacific Basin offered an opportunity. Such a development, however, required greater reciprocity in trade matters and a less hostile attitude to Asians in general. Thus when Japan was admitted to GATT in 1955, Australia quietly began negotiations which led to an agreement on commerce whereby Japan and Australia granted each other 'most favoured nation' status. Gradually, too, some of the discriminatory legislation against Asian immigrants was removed during the late 1950s and 1960s and finally in 1973 discrimination against potential migrants on the grounds of race was replaced by selection on the basis of skill and the reunion of families.

Predictably the 1957 trade agreement with Japan created a furor. On the one hand, primary producers—who accounted for some 80 percent of Australia's exports but employed about 8 percent of its work force—welcomed the possibility of new export markets. On the other, manufacturers—who employed some 30 percent of the work force—and trade unions were concerned that there would be a flood of cheap goods which would erode Australian industry. In fact the agreement with Japan was hedged about with safeguards and escape clauses relating to "unfair" competition and so the political conflict between farming and manufacturing lobbies was deferred. Nonetheless, the dilemma remains of how to balance the priorities and privileges to be granted to those sectors that create export income but relatively few jobs as against those that provide jobs but produce relatively little export income.

During the 1970s and 1980s policy developments in Australia failed to create a coherent set of strategies appropriate for the changing international economy. The reasons for this failure are complex but can be summarized briefly. First, the so-called mineral boom during the second half of the 1960s, along with the international fixed exchange rate system that applied at the time, led to a trade imbalance and a surge of imports, especially of consumer durables, which were much cheaper and arguably of better quality than the locally made equivalents. Australian manufacturers sought further protection and subsidies rather than taking positive action to improve their efficiency, modernize plants, equipment and products. Neither manufacturers nor governments attempted to rationalize the grossly distorted structure of industry that had arisen because of piecemeal application of, and increases in, tariffs over many decades; the competitive activities of the states; a vast array of labour market and institutional rigidities; and the antipathy by a succession of conservative federal governments to the establishment of an advisory body concerned with long-term economic strategies or to an overhaul of the ad hoc practices that had developed. Instead, monetary and fiscal measures were relied upon to influence the private sector as a whole and the inequities and inefficiencies of the existing structure of activity were largely ignored.

Second, in December 1972 a Labor federal government was voted into office, the first such administration for over two decades. Although it remained in office only until November 1975 it generated considerable controversy then and in hindsight. Basically it aimed to bustle Australia into the twentieth century by reforming almost overnight what it perceived to be long-standing archaic or inadequate policies relating to social security, health, education, housing, urban and regional planning, the status of women, and so on. The result was a flurry of measures which, among other things, raised direct and indirect wage costs at a rate that exceeded those of most of Australia's trading partners. For instance, during the five years to mid-1975 minimium hourly wage rates in Australia rose by 120 percent compared with rises of 43 percent in the United States and 63 percent in West Germany. Whatever the long-term judgment on these reforms may be, the point for present purposes is that they further eroded Australia's ability to compete against imports or make inroads into new external markets.

Third, in May 1975 Australian economic and social planners, and business confidence in general, were shattered by an authoritative report which argued that instead of the much-vaunted population prediction of 28,000,000 by the year 2000, the more likely figure, after allowing for immigration, would be only 18,000,000 of which a significant proportion would be dependent on pensions and other old-age benefits. The accuracy of the report may be judged from the fact that the present official estimate is for a population of 18,000,000 in mid-1993 and 24,300,000 in 2020 (Australia, Department of Immigration,

Local Government and Ethnic Affairs, 1988). Part of the recent increase in the estimate is explained by Australia's more than token response to the plight of refugees from Vietnam and other Asian countries and, more generally, by a growing awareness that an increased flow of immigrants is necessary to offset the declining rate of natural increase and the "greying" of the population. In 1989 about 13 percent of the population was of pensionable age: by 2020 this is expected to rise to 25 percent.

The downturn in global economic activity began to be felt in Australia in 1974 following the breakdown of the Bretton Woods international monetary arrangements that relied on fixed exchange rates, the activities of the OPEC cartel, concern about "limits to growth", and the restraints on world trade in both products and services. To some extent, however, the full effects were cushioned until the early 1980s by the high level of exports of mineral and farm products to several Asian nations which continued to have high rates of growth. Even so, the problems facing Australian manufacturing already seemed obvious enough to a government committee established to advise on policies for industry, which reported in October 1975 (Australia, Report ... by the Committee to Advise on Policies for Manufacturing Industry, 1975). Under the heading (p. 1), "Deep-seated malaise," this Committee noted:

> Relative to other countries, Australian industry was built in a remarkably short period. Dedicated and energetic people responded to challenges and opportunities. But Australian manufacturing was largely created to serve a growing domestic market by deliberate policies of import substitution, immigration, fixed exchange rates and capital inflow. Growth helped industry cope with inflation and other problems. Now that the domestic market is satiated and can grow only slowly, most manufacturing is stalled and lacks purpose. It needs exports to grow. But the industry structure created by those earlier policies is not well suited to the challenges of international competition. And just as industry has stalled, increased pressures have been placed on it. Prices and productivity have not increased as fast as wages; industrial unrest has increased; protection has been reduced; the dollar has been revalued; credit has been squeezed; export opportunities have been reduced. Much of Australian manufacturing sees the combination as so overwhelming as to threaten survival. Some see a

> socialist plot to destroy them. The breakdown of confidence is general.

By the mid-1980s little had changed within Australia to improve its position in the global economy. Although the new Labor federal government elected early in 1983 was soon able to negotiate a wage restraint accord with the unions, which has proved to be relatively successful in containing direct wage costs, it has only gradually been able to tackle other fundamental problems such as low productivity, lack of competitiveness, labour market rigidities, mounting budget deficits, tariff and other forms of import protection, and the national dependence on exports of pastoral, farm and mineral products.

Indeed, in the mid-1980s Australia was facing its most-serious ever economic crisis. This was recognized in no uncertain terms in May 1986 when the Federal Treasurer publicly warned that unless people in Australia woke up to the problems facing their country they would see it slipping very quickly into the status of yet another "banana republic". This was a brave, if potentially dangerous, statement. However, instead of its being seen by the international community as an admission of weakness, it was regarded as a positive sign that, at long last, Australia had not only recognized, but was prepared to tackle, its basic long-term economic problems. It also had the effect of putting a somewhat stunned Australia into a "reconstruction mode" with a growing awareness of the need for greater flexibility in addressing the issues involved.

As if to reinforce this, the worldwide stock market crash in October 1987 brought home to people Australia's vulnerability to international events over which it has little or no influence: one estimate, for instance, is that 17 of the 50 companies in the world worst affected by the crash were Australian (MacDonald, 1987). More generally, Australia has almost no control over a wide range of uncertainties like the rate of growth of the world economy, exchange rates, the terms of trade (the ratio of import prices to export prices), interest rates, trade barriers, food growing and marketing subsidies, and the budgetary policies of—in particular—the United States and Japan.

Even if scenarios such as a world recession and a resurgence of protectionism turn out to be too pessimistic, Australia faces a deteriorating external environment. In 1980 net foreign indebtedness was 6 percent of GDP; by September 1989 it stood at A$110,000 million, or 32 percent of GDP. The 1989 current account deficit was A$21,000 million compared with A$13,800 million in calendar year 1988. The cost of servicing this debt (interest and capital repayments) puts Australia on an endless treadmill, as it is likely to absorb an ever-increasing proportion of export earnings unless this debt can be reduced. Ironically, this problem has been exacerbated by decisions in 1984 and 1985 to deregulate the Australian banking system which reshaped retail and commercial banking; moreover the removal of controls on financial flows revolutionized corporate funding. Offshore capital markets became accessible to a much wider range of business, and corporate debt rose so that at the end of September 1989 almost half Australia's net foreign indebtedness was owed by the country's 50 largest enterprises. Argument continues as to how much represents investment in property including facilities which may or may not attract overseas tourists; and how much represents the maintenance of an artificially high and—in the long-run—unsustainable standard of living.

The need to boost exports helps to explain why Australia has been pushing in recent GATT negotiations for freer world trade in goods and services and the abandonment of practices (especially in the US, the EEC and Japan) which lead to agricultural surpluses, subsidized exports of foodstuffs and the erection of quotas and other barriers to imports. Agricultural subsidies, of course, do not threaten the Australian economy alone, but also those, for example, of Canada, New Zealand and Argentina.

The central issue facing Australia is that in order to achieve a substantial increase in exports there will have to be a substantial increase in the penetration of imports. As McGuinness (1987) has observed, it is the excess of the former over the latter which is the key to reducing overall indebtedness, not the reduction in import penetration. This requires the development of additional exports of goods and services (like medical equipment and higher education), and also the reduction of costs (especially wages and wage-related costs) and levels of industry protection to make existing products more competitive on both the international and domestic markets.

The possibility of tackling some of these long-standing issues was enhanced by the decision in December 1983 to float the Australian dollar. Because the Australian economy then became subject to external monitoring, the federal government was enabled and compelled to exert influence as "responsible economic managers" over the attitudes of the business community and the unions. The series of collapses by the Australian dollar from early 1985 to mid-1986 quickly forced a new discipline on economic policy, including a break from automatic wage indexation, the removal of the 13.5

percent interest ceiling on new mortgages, and moves towards a balanced federal budget. Manufacturers began to accept the need for greater rationalization and lower rates of protection; the Australian Council of Trade Unions grudgingly accepted wage increases that lagged behind the rate of inflation; and a mixture of cuts in expenditure, the more efficient collection of taxation revenue and the sale of assets, enabled the federal government to achieve a domestic surplus of A$580 million in 1987-88 and nearly A$5,900 million in 1988-1989 compared with a series of deficits that peaked at A$7,960 million in 1983-84. If surpluses on the domestic account can be maintained this will help stabilize the foreign debt because Australia will have less need to call on overseas savings to make up for the shortfall of domestic savings required for investment. In 1988-89 retirement of federal overseas debt amounted to A$3,200 million; in 1989-90 it is expected to fall by another A$1,750 million. The other side of the coin, however, is that the federal government has been obliged to maintain very high interest rates by international standards (over 17.5 percent for new house mortgages) to try to take the fire out of Australia's overheated economy. A survey of the 1980s, appropriately entitled "zany optimism yields to sober reality", by the Australian Financial R*eview* (December 11, 1989, p. 14) concluded:

> while it has not been, as some argue, an entirely gloomy period for Australia, the contrasts between the zany optimism of the resources boom of the early 1980s and the nation's tenuous grip on its present standard of living as it enters the nineties could not be more stark.

The social implications of this are considered later in the chapter.

3.3. SPATIAL PATTERNS

In broad terms the spatial framework of the Australian economy had been established prior to Federation in 1901. The boundaries of the colonies, most of which followed lines of latitude and longitude or river courses, became the boundaries of the states. This perpetuated the situation in which transport systems channelled produce to the capital city concerned, rather than to a nearer major port. Activities associated with the import and export trade, manufacturing, commerce and higher-level services reinforced the capital cities of the five mainland states which together contain over 60 percent of the population. The other seven cities with more than 100,000 inhabitants (of which only Canberra is located inland) accounted for a further 10 percent.

By the end of the nineteenth century, the climatic limits to farming and pastoral activities had been established by a process of trial and error. The subsequent introduction of new strains of seed and breeds of cattle made only marginal differences. In effect, the observation of Griffith Taylor (1951) that "future millions of Australians are going to find their dwelling places and occupations in the lands already known by 1865" has proved to be wellfounded. Urban settlement has continued to cling to narrow coastal strips, little more than 200 km wide, along the east, southeast and southwest coasts. In this so-called 'closely-settled zone' which makes up only 3.3 percent (253,500 km^2) of the total land area, live over 80 percent of the population.

Topography, climate, relatively low heat discomfort, reliability of water supply, and people's desire to be near sea and surf have all been influential, as has the fact that Australia has no commercially navigable river or inland waterway. Plans sponsored in the early 1970s to promote relatively modest development centres inland fizzled out in less than a decade. These had been seen as a counter to the ever-growing dominance of capital cities and as a way of providing an alternative lifestyle to the frenetic, polluted and commuter-orientated existence of people in Sydney and Melbourne. The success of these growth centres (at one stage no fewer than eight were proposed) was predicated on their quickly reaching a self-sustaining critical mass by the transfer of federal and state public servants and the location there of manufacturing. No sooner had these plans been formulated than the public service was cut back in the interests of "small government," and manufacturers were being pressured to reorganize their operations to achieve economies of scale and greater efficiency. Moreover, young professionals wanted to remain close to established educational and other services for themselves and their children, while people nearing retirement preferred to invest their life savings in property along the coast. The rate of growth of Australia's largest cities has now slowed because of a decline in the rate of natural increase and the migration of older people to resorts and retirement villages. To some extent these trends have been compensated for by the propensity of overseas immigrants to concentrate in the major cities where the various ethnic groups have established their social and cultural associations and activities.

The only significant change to Australia's settlement pattern was the construction during the 1960s and 1970s of some 30 small mining and mineral processing towns in isolated parts of the continent. Along with the associated railways, power-generating plants and port facilities, these were for the most part paid for and run by the mining companies themselves. However, this form of development has had its day since such specific-use infrastructural investments make it harder for miners to respond to global market changes. In future, with only a few exceptions, companies will seek to establish new operations that can be serviced by existing residential, commercial and transport infrastructure or which can be supported by long-distance aerial commuting services.

While urban places along the coast are booming, many of the long-established inland country towns, often originally spaced according to the distance a bullock-dray laden with wool bales could travel in a day along roads that were little more than tracks cut through the bush, are in decline. The construction of high-speed motorways that usually by-pass settlements, the rationalization of hospitals, education and custodial facilities, and the closure of farm-related services like wool-selling and stock and station agencies, have all had a domino effect. Places which a few years ago complained that they lived in the shadow of a prison or psychiatric hospital now petition for their retention. The slashing of public-sector spending by the federal and state governments (which has already been mentioned) and the rationalization by the private sector means that this thinning-out process will continue, despite the electoral backlash discussed later that it engenders. Government support in the 1970s for alternative-lifestyle rural communes is now seen as a reaction rather than a reality.

Gone, too, are most of the grandiose plans that may have made some impression on the face of Australia. Thus, irrigation schemes in the north have proved to be expensive white elephants, and plans to reverse the flow of some rivers in Queensland to irrigate inland areas are quickly gathering dust. Nonetheless, some projects remain. The feasibility of building a 1,300 km railway from Darwin to Alice Springs is one, and the possibility of linking Sydney, Canberra and Melbourne with a new 700-km high-speed rail link is another. In general, however, most of the activity in Australia in the foreseeable future will be confined to the cities and the coast.

3.4 ATTITUDES AND IMPACTS

In the 1960s, Australia was dubbed the "lucky country" in which a spirit of "mateship" had long prevailed. Arguably neither description is true. In fact Australia was lulled into believing that the opening up of coal, iron ore and bauxite resources in the 1960s and 1970s would be a long-running gravytrain. Even as late as 1980, during the federal election campaign, the incumbent

Conservative government made much of the boom likely to flow from projects associated with the production and processing of minerals. Many such schemes, however, were predicated on naive assumptions about the growth of the global economy. For example, selective reports from federal departments in 1980 suggested that Australia's aluminium producing capacity would expand from 367 million tonnes that year to 1,600 million tonnes in the mid-1980s: in fact installed capacity stood at only 853 million tonnes at the beginning of 1986 and 1,120 million tonnes early in 1989. Many similar examples of promoting unrealistically optimistic projections for political ends could be cited.

Australian society, however, was already becoming divided between the "haves" and the "have nots". On the one hand, organized labour is very powerful with 310 or so separate unions accounting for 55 percent of all wage and salary earners. It has been noted already that the Australian work force gained very significant wage increases and other benefits during the 1970s and early 1980s. The federal government and the more responsible trade union leaders found it difficult to persuade people that Australia had shifted from being a lucky country to one which was in serious economic difficulty and where living standards and other expectations would have to be eroded. Until recently the union movement had been reluctant to give up many of the perks it had gained, although during 1987 it began to do so as wage increases were in part tied to the removal of restrictive trade practices that reduce productivity.

Of particular potential importance have been actions commenced under the provision of the 1989 (federal) *Industrial Relations Act* which allow enterprise agreements and reduced union coverage. For instance, at one oil refinery in Sydney there are 23 unions; a dispute by a dozen or so employees can bring the whole operation to a standstill. Amalgamation is supported by the union leadership which wants to carve Australia's messy craft union structure into 20 or so neat industry federations, but it has yet to gain acceptance by hard-liners in the smaller unions who have been used to fighting what they term "body-snatching" by one union from another.

On the other hand, the unemployed, who numbered 491,000 in November 1989 (six percent of the work force in seasonally adjusted terms), have little power and few perks. Moreover, there are estimated to be a further 750,000 people who, while not officially recorded as out of work, were still looking for a job or had become "discouraged job-seekers". Even so, there is a marked reluctance by organized labour to allow any redistribution of job opportunities, and hence income, to the unemployed. Thus teenage unemployment remains high with many youngsters being forced to take on part-time "dead-end" jobs (such as counter-hands in fast food outlets) which are unlikely to lead to a career or even make them eligible for permanent work. Some employers argue that because there is so little difference between wages and wage-related costs associated with teenagers and adults, there is little incentive for them to take on inexperienced school-leavers. This sets up a vicious circle from which some youngsters see no hope of escaping. Unions have consistently opposed suggestions that teenage wagelevels should be reduced because they see this as the thin end of the wedge that might shatter existing remuneration packages. The unions have used their bargaining power over wages to extend what is called "occupational superannuation" from about 50 percent of the work force in 1980 to over 80 percent in 1990—a process encouraged by the federal government, largely through favourable tax treatment. While, these superannuation innovations may increase the pool of domestic savings available for industrial development and thus decrease the need for overseas borrowing, they are likely to transform attitudes to welfare payments with an inevitable contraction of benefits from the public purse. The gap between the"haves" and the "have nots" is thus likely to widen.

There is considerable ambivalence, too, about Australia's immigration program. Since the Second World War, Australia has admitted some 4,600,000 immigrants as permanent residents (that is, settler arrivals). The size of annual intakes of settler arrivals has fluctuated considerably during this period, ranging from 186,000 in 1969-70. to 53,000 in 1975-76. The 1987-88 total was 143,500. Of particular concern to some people is the fact that migrants from Asia and the Middle East made up only 30 percent of the total arrivals in 1977-78 but 41 per cent in 1987-88. The official view is that without immigration Australia's population will begin to decline by the year 2030, assuming no further fall in fertility and that, as migrants tend to be in the younger age groups they will help to retard the ageing of the population and the aged dependency ratio. Others, including those who came to Australia during previous immigration waves (such as those from southern Europe), see migrants as competitors for their jobs and have yet to be convinced that the growth of population from this source means new household formation and therefore greater demand for goods and services. However, they see migrants, many of whom have only limited oral and written English ability, as useful for doing the kinds of jobs in car assembly plants and steelworks that no one else wants. People who raise issues about the way the immigration program operates (such as perceived Asian biases in the family reunion program) and the rate at which a multicultural society can develop without creating social tensions are pilloried and branded as racist.

Enough has been said to indicate that Australia is facing a series of economic and social tensions as a result, in large measure, of its belated and hence rapid attempt to internationalize its outlook. However, century-old attitudes cannot be changed overnight, and this dilemma is particularly evident in a country where there are over 800 state, territory and federal politicians serving an electorate of barely nine million, and where few governments serve out their three- or four-year terms. Adding to the "juggling" involved is the desire of the federal administration to try to ensure that state governments with similar political philosophies remain in office or are returned to power. A particular example of the difficulties facing politicians who espouse moves to make Australia competitive at an aggregate macro-level is that many electorates are finely balanced between political parties so that a handful of votes can make all the difference as to which of them is able to form a government. For instance, there are a dozen or so country towns in New South Wales and Victoria in which between 25 and 60 percent of the labour force works in the textile, clothing or footwear industries, activities which in 1988-89 were protected by effective tariffs of 74 percent (textiles), 158 percent (clothing) and 218 percent (footwear). Although the number of such places and the total of jobs at risk (perhaps 25,000) are small, neither the federal government nor the state governments concerned can afford an electoral backlash in the constituencies involved. Thus decisiveness and strong economic management have to be tempered with the ever-present reality that politicians usually depend on the swing of a handful of votes in electorates with relatively small electoral rolls where local issues are never far from the surface. The division of responsibilities inherent in a federal system complicates this problem but confusion over aims, policies, mechanisms and time-scales is compounded by the truly byzantine internal politics of the main parties in Australia which effectively submerge long-term coherent objectives beneath the daily froth and bubble of internecine squabbles.

For decades Australia sought to insulate itself from the market mechanisms of the global economy. As a result, an artificial structure was created which, almost overnight, is being demolished in the face of, for example, a deregulation of the financial system and the privatization of government monopolies and statutory authorities. It can also be argued that the current plans to reorganize the public service, of higher education and research institutions, and of the taxation system, are long overdue and will be beneficial in the longrun. In the meantime there has been a loss of confidence. People in their twenties and thirties have learned that careers can crash within a few hours when the stock markets and currency markets react adversely, and academics, research workers and public servants are being enticed to resign in their mid- to late fifties by the offer of reformulated and more attractive "early retirement packages" which in some instances are little more than "redundancy payouts."

The long-term prospect then is the continued growth of "two Australias": in simplistic terms, those with work will continue to become relatively better off while those without work will become more numerous and relatively poorer. Teenagers are becoming increasingly disillusioned by work programs that simply recycle them back on the dole after six months because employers are no longer eligible to claim wage subsidies. Furthermore, schemes designed to improve basic job qualifications, such as teaching migrants to communicate, are becoming victims of public-sector funding cuts. A study at the University of Melbourne has estimated that three million people are living below the poverty line, by which is meant the amount required to pay for basic necessities.

The prospects are gloomy and it is hard to see remedial measures that would be workable in reality given the dogmatic stances and institutional rigidities. Australia faces the possibility that its per capita income (admittedly not always the best measure) may soon fall behind that of several Pacific Basin countries. In 1987 Japan's per capita income was US$15,760 compared with Australia's US$11,100: next in line were Hong Kong (US$8,070) and Singapore (US$7,940). While the economies of several developing countries in southeast Asia suffered setbacks in 1985 and 1986, the prospect remains that countries like Singapore, South Korea and Taiwan could overtake Australia during the next 10 to 20 years. "No Australian government," one commentator has suggested, "could survive the national embarrassment." The leader of a Japanese trade mission summed up the situation when he commented, "Australia has become accustomed to lounging in a luxury hotel swimming pool: outside the pool is the hard, cruel sea".

ACKNOWLEDGMENTS

The author is grateful to Peter Wilde and Jan Linge for their advice and practical assistance.

4. CANADA'S INTERNAL CORE-PERIPHERY STRUCTURE

David Walker

This chapter reviews elements of Canada's regional structure, examines recent trends and changes concerning Canada's space-economy, and contributes to the debate on appropriate economic and regional policy for the country. Reference is made particularly to employment change, migration patterns, and income levels, which are variables traditionally used in assessing interprovincial performance. In addition, a consideration of ownership changes and the operations Canada's corporate elite is offered. The conceptual framework underlying this discussion is Friedmann's (1972) view of core-periphery relations, which stresses the element of domination and control by a core over a periphery.

Canada is a large country that grew as a series of independent colonies until confederation in 1867. Although the country developed a sense of unity in the succeeding years and was forged together by the National Policy, the process of integration, particularly the incorporation of Atlantic Canada into the economic mainstream of Canadian life was slow. A national corporate elite finally emerged in the early twentieth century. Subsequently, the largely western parts of the country were settled and developed under the strong influence of southern Quebec and Ontario. Thus at an early stage in Canadian economic history a national scale core-periphery pattern had emerged with the core centring on the Quebec-Windsor corridor and dominated by Montreal and Toronto. The corridor itself is highly differentiated in function and character and recent trends have seen its centre move from Montreal to Toronto (McCann, 1981, pp. 2-35).

This chapter concentrates on the internal core-periphery relations with a focus on the provincial level, because data availability is so much better at this scale. In practice, of course, Quebec and Ontario have large peripheries themselves and their northern sections are more akin to other parts of the country than to the southern parts of their respective provinces. It may also be argued that such cities as Calgary, Edmonton and especially Vancouver have achieved metropolitan status since for example, they contain head offices of national importance. At another scale, it is widely supposed that Canada as a country is a periphery in the world at large. While other studies have explored this view (Britton and Gilmour, 1978; Galois and Mabin, 1981) the present chapter is limited to a description and discussion of broad provincial trends.

4.1 RECENT REGIONAL TRENDS IN CANADA

Overall, Canada's population has grown steadily in the last 20 years but with a markedly faster rate in the west than the east (Table 4.1). The rapid expansion of Alberta and British Columbia was particularly notable in the 1970s and continued an established pattern. The two provinces in 1951 had only 15 percent of Canada's population. Now they have around 21 percent. It is not yet clear whether Alberta's growth will slow down. The country west of Ontario, however, has not been one of rapid growth everywhere. With 8.2 percent of total population now, in contrast to 8.9 percent in 1951, Saskatchewan and Manitoba have both lost a little in relative terms. Meanwhile, not only has Atlantic Canada continued to slip in importance, but it has been joined by Quebec whose relative decline since 1951 has been greater. Ontario has maintained its share of the Canadian population. So it is clear that the western end of the country has gained strength.

The 1970s was a decade of rapid employment growth which was nowhere more evident than in Alberta and British Columbia (Table 4.2). The boom in oil and related activities particularly helped Alberta while British Columbia's success was more broadly based. During the postwar period, unemployment has been especially low in the Prairies until the recession of the 1980s. More recently, low oil prices, and bad harvests have re-emphasized the Prairies' dependence on primary activities and their failure to breakfrom the mould of a staple economy. British Columbia's unemployment, meanwhile, always remained higher and in the 1980s almost became comparable with the Atlantic Region. There were fears, that once the World Exposition of 1986 was over the province would relapse back into the severe recession which began in 1981. Atlantic Canada has not succeeded in improving a weak economy to take up the slack in employment, while Quebec's unemployment situation has also worsened noticeably, despite a recent strengthening of the economy. Thus Canada's unemployment looks bleak and there is concern that the economy is not capable of rebuilding to take up the slack. Even recent growth in Southern Ontario has barely been sufficient to bring unemployment levels back to previous lows. It is recognized that primary production in Canada faces much competition in world markets. Meanwhile Canada's manufacturing sector still has many old, outdated operations and too great a concentration on mature industries.

TABLE 4.1

CANADA: POPULATION, BY PROVINCE (000s)

	1971		1981		1986	
	NO.	%	NO.	%	NO.	%
Newfoundland	522	2.4	568	2.3	5568	2.2
P.E.I.	111	0.5	123	0.5	127	0.5
Nova Scotia	789	3.7	848	3.5	873	3.4
Quebec	6026	27.9	6438	26.4	6540	25.8
Ontario	7703	35.7	8625	35.4	9114	35.9
Manitoba	988	4.6	1026	4.2	1071	4.2
Saskatchewan	926	4.3	968	4.;0	1010	4.0
Alberta	1628	7.6	2237	9.2	2375	9.4
B.C.	2185	10.1	2764	11.3	2889	11.4
Yukon	18	0.1	23	0.1	24	0.1
N.W.T.	35	0.2	46	0.2	52	0.2
CANADA	**21568**	**100**	**24352**	**100**	**2535**	**100**

TABLE 4.2

THE EMPLOYMENT PICTURE IN JANUARY (000s)

	Employed				
	1972	1975	1980	1985	1987
Atlantic	575	708	743	792	782
Quebec	2133	2455	2569	2631	2789
Ontario	3096	3520	3931	4178	4506
Prairie	1281	1531	1827	2201	2035
B.C.	832	1005	1123	1168	1221
CANADA	**7917**	**9227**	**10606**	**11117**	**11545**

	Unemployment Rates				
	1972	1975	1980	1985	1987
Atlantic	12.9	10.6	14.3	18.2	16.9
Quebec	9.8	8.1	10.5	13.0	11.7
Ontario	5.8	6.0	7.6	9.9	7.4
Prairie	5.8	3.2	5.3	10.0	10.3
B.C.	8.7	7.5	9.6	16.4	15.3
CANADA	**7.7**	**6.7**	**7.4**	**11.2**	**9.3**

Source: CAT 71-001 The Labour Force

TABLE 4.3

CANADA'S LABOUR FORCE BY SECTOR, 1986

	ATL %	QUE %	ONT. %	PRAIRIES %	B.C. %	TOTAL %
Agriculture	2.2	2.5	2.5	9.4	2.6	3.7
Other Primary	5.7	1.5	0.0	4.5	4.7	2.6
Manufacturing	12.6	19.6	22.6	8.3	11.0	17.1
Construction	6.7	5.2	6.0	6.1	6.4	5.9
Transport, Communications & Utilities	8.0	7.3	6.7	7.7	8.6	7.3
Trade	19.7	17.8	17.6	18.8	18.8	18.1
Finance, Ins. & Real Estate	3.7	5.3	6.3	5.2	5.9	5.6
Services	32.2	33.1	31.6	3.0	35.3	32.6
Public Admin.	9.2	6.6	6.3	7.0	5.9	6.6
Other	0.0	1.1	0.4	0.0	0.8	0.5
TOTAL (000s)	**939**	**3189**	**4668**	**2270**	**1451**	**12717**

Source: Statistics Canada, Cat. 71.001

TABLE 4.4

PERSONAL INCOME PER CAPITA

	1970		1975		1980		1984	
	$	%	$	%	$	%	$	%
Newfoundland	1983	63.4	4115	68.6	6643	75.3	9702	67.3
P.E.I.	2082	66.5	4214	70.2	7187	70.6	10310	71.6
N.S.	2423	77.4	4949	79.1	8032	78.9	11693	81.1
New Brunswick	2252	71.9	4632	77.2	7411	72.8	10734	64.7
Quebec	2774	88.9	5470	91.2	9586	94.2	13487	93.6
Ontario	3705	118.4	6596	109.9	10953	106.7	15841	109.9
Manitoba	2906	92.6	5784	96.4	9033	88.8	13743	95.4
Saskatchewan	2267	72.7	6240	104.0	9271	91.1	13006	90.2
Alberta	3105	99.2	6182	103.0	11107	109.1	15376	106.7
B.C.	3405	108.5	6489	108.1	11348	111.5	14778	108.3
Territories	2960	94.6	6297	104.9	10328	101.5	15611	108.0
CANADA	**3129**	**100.0**	**6001**	**100.0**	**10178**	**100.0**	**14412**	**100.0**

Source: Statistics Canada, Cat. 13-201 National Income and Expenditure Accounts

The breakdown of the labour force shows some major contrasts within the country, especially between core and periphery (Table 4.3). Ontario and, to a lesser extent, Quebec have a strong manufacturing work force, whereas other parts of the country lag many percentage points behind. In contrast, primary activities come out strongly elsewhere, with agriculture dominating the Prairies and other primary activities, including primary manufacturing, in the remaining regions. Of course, in reality, both Quebec and Ontario have large primary-oriented peripheries of their own so that the prime manufacturing area is spatially very concentrated. The peripheries still suffer from wide fluctuations in economic fortunes, and no amount of activity outside manufacturing seems to be able to break this disadvantage. Most other occupations are fairly comparable across the country except that the central provinces employ a smaller proportion of their work force in public administration. This characteristic mainly reflects the importance of indivisibilities in providing public sector services.

Perhaps, the best overall picture of provincial development comes from data on personal incomes (Table 4.4). In general, the provincial figures have been moving to a national norm. The very low levels in Atlantic Canada, which incidentally were nearly a further 10 percent lower in 1951, have been brought closer to the national average. Quebec has also moved up, while Ontario and British Columbia have moved down towards the Canadian norm. Albertan per capita incomes increased rapidly during the 1970s while the per capita incomes of Manitoba and Saskatchewan (especially the latter) have tended to fluctuate with crop results. In general, trends were in the right direction until about 1981. Since then, however, there have been some signs of a reversal in equalization trends as per capita incomes in Ontario have increased faster than the national average. Unfortunately, much of this balancing has been achieved by welfare programs rather than by productive activities. Many feel we have only succeeded by building up huge public debts, amounting in 1982 to C$38.9 billion (Financial Times, July 5, 1982). Thus, the view has grown that we must get back to basics and be less concerned about spatial impacts (Courchene and Melvin, 1986).

Changes in business conditions have encouraged considerable spatial mobility of the labour force. According to the 1981 census, the eastern part of the country showed net losses from internal migration between 1976 and 1981: east of Ontario only Prince Edward Island had a small positive balance. Newfoundland's loss of over 14,000 was quite significant for a small province but Quebec's absolute total (42,000) was larger. In the west, only

Manitoba lost. The really big winner was Alberta with over 200,000 net migrants to its work force. British Columbia also gained substantially (116,655), while Ontario's total was lower (71,000). These movements of course helped balance the population and the new job situation, and most migrants were in management, construction, clerical or manufacturing jobs. Interestingly, Ontario was both the largest supplier and receiver of migrants to the labour force, illustrating the considerable cross-movements that take place.

Four provinces (Quebec, Ontario, Manitoba, British Columbia) received more immigrant workers than workers from any Canadian province. Saskatchewan received more workers from Ontario than any other part of the country, and for Saskatchewan it was second to neighbouring Alberta. The second sources of supply, however, reveal an east-west split with migrants remaining either east or west of Ontario. The implication of Quebec's linguistic isolation is perhaps manifest in its relatively high dependence on immigrants. Of course, this was a period of strong Quebecois "nationalistic" feeling. Substantial numbers moved away.

Canadian society is reasonably mobile. Given the expense and inconvenience of movement, it is remarkable that about one third of the 1981 workforce had moved in the previous five years, and that roughly one tenth had left province or country. Can one expect more adjustment without some form of financing for movement? We probably could use a much closer look at why people do not move, in terms of financial and other costs of movement. To expect more migration without care about the migration process, seems to be courting disaster.

To close this section, it should be noted that both British Columbia and Alberta have experienced net out-migration in recent years. A recent report (MacLean's, 1986a) notes that BC had a net loss of 7,000 via migration in 1985. It was also reported that 96 percent of net new jobs in the first four months of 1986 were created in Ontario, while Alberta had a net loss of 11,000 jobs. The article talked about a possible "massive depopulation of Alberta over the next two years" (Financial Times, May 19, 1986, p. 8).

4.2 BUSINESS CONCENTRATION

In terms of population and per capita income, therefore, there has been a trend in Canada for wealth to be spread somewhat more fairly across the country. But core-periphery relations are also about control, and in the realm of business, control centres have remained concentrated. It must also be noted that much ownership lies completely outside the country.

The extreme dominance of Ontario as a headquarters for so called "industrials," including a wide range of business other than financial institutions remains. It controls over half of the top 200 of these businesses. Admittedly, this proportion has been declining, from almost 60 percent in 1975 to 51 percent in 1985. Montreal, which was once more important than Toronto, still retains some strength but in the late 1970s and early 1980s some national companies moved away. Higgins (1986, pp. 66-69) lists 42 enterprises which moved headquarters or administrative units from Montreal in this period, all but one to Ontario. In western Canada, there are still not very many large locally owned businesses. Nevertheless, by 1985 they totalled 30 percent of headquarters of the top 200, and the Prairies even had five of the top 20. These companies are strongly concentrated in the resource and related sectors sectors but the percentage in the top 200 has almost doubled since 1975. The Financial Post 500 shows, however, that central Canada has retained a stronger position in many key service sectors (23/25 life insurance head offices; 14/15 property and casualty insurance; 13/15 investment companies; 7/10 real estate businesses). Where the traditional corporate elite is at its strongest, the core retention has been very high.

For the industrials, it is remarkable how many Ontario-based companies are foreign subsidiaries. In fact, despite a general tendency for foreign ownership to decline in Canada, it has continued to rise in Ontario. Despite some periods of strong attack against foreign investment and a period of official federal controls designed to restrict it, there has always been strong support in favour of foreign investment. At present, the desire for free trade is dominant in federal circles, following a view that efficient operation is what Canada needs, and increased competition will help it to come about. After years of equivocation, we may even get a new competition act in effect. There has been a long buildup to the current pattern of control (Clement, 1977) and it will not be quickly changed. In fact, modifications of core management techniques may offer the best bet. Some multinationals are decentralizing (Toffler, 1985) and a few have established world product mandates in Canada. This implies that very considerable decision-making powers remain in the country, subject to Canadian influences (Science Council of Canada, 1980).

There remains in Canada a remarkable concentration of economic power in the hands of a few individuals and families. There are a few real industrialists, such as the McCains of New Brunswick, with a french fries empire that extends to many countries, but most have their base in the wheeling and dealing of real estate and finance. The operations of this group have been very much to the fore recently as they have been exposed in a series of mergers, takeovers, and takeover attempts. The traditional elite is very much a Toronto/Montreal group, with its own educational institutions and clubs (Clement, 1975), and its own companies which operate all over the country. To take two examples; the Montreal- based Bronfmans control the massive Seagram's business worldwide and are in charge of a major real estate company, Cadillac Fairview. In addition, they have substantial influence in E.I. DuPont de Nemours, as well as in footwear (Warrington Inc.) and oil, gas, coal and related activities (Bow Valley Industries, Campbell Resources). Their assets amount to about C$36 billion. As a second example, Kenneth Thompson (son of Lord Thompson of Fleet) returned to Canada with major profits from North Sea oil, and established a massive retailing business (The Bay, Simpsons, Zellers) to complement ownership of international and Canadian newspapers, and oil and travel interests. Assets are around C$13 billion (The Financial Post 500, summer 1986).

Changes in the regional importance of top companies have taken place partly due to the relative growth rates and the rise of new business, and partly, because some companies have moved headquarters. I referred earlier to companies leaving Montreal, but there has also been a movement of oil companies to Alberta. For example, Shell Canada and Gulf Canada (Nos. 9 and 10 in 1986 by sales) with headquarters in Calgary, were both based in Toronto in 1981. The federally owned Petro-Canada, a relatively new operation of increasing size, is also based in Calgary. On the other hand, takeovers and mergers have played a considerable role as well. Petrofina, another oil company which was based in Montreal, was taken over by Petro-Canada and lost its identity. The historic Hudson's Bay Company was taken over by Kenneth Thompson and its head office was shifted from Winnipeg to Toronto.

Merger movements have had a major centralizing influence and, for example, were important in concentrating head offices in Toronto and Montreal earlier in the century. Present trends are unclear. A substantial number of new entrepreneurs has become established in the country. Newman (1981) has characterized them as the "acquisitors" and noted substantial differences in their behaviour from that of the traditional corporate elite. While some of the most successful are based in Toronto, such as the Reichmanns who immigrated to Canada after the Second World War, others are westerners. On the whole, this latter group has been keen to stay out west and to build up the regional economy. They have been supported by some local politicians, especially by the former premier of Alberta, Peter Lougheed. In some instances, there is a question as to the viability of their businesses. In the difficult times of the last few years, it is clear that not all the operations which seemed sound in the boom years were built to last. Even two banks have folded in Alberta, something unknown to recent generations of Canadians. With the future still looking bleak on the oil patch, it may be that more companies will be sold or snapped up by eastern predators.

I believe that a substantial number of western entrepreneurs will weather the storm. However, there is a new generation of entrepreneurs that has arisen in Quebec under Parti Quebecois encouragement. Most are still small but the renewed business interest there is leading to a growth of indigenous enterprise (McLeans, August 4, 1986b). Taken together, these two groups may well provide the backbone for resistance to the increasing buildup of a Toronto-based core in the country, but we cannot be sure for another 20 years at least.

4.3 CONCLUSION

In assessing Canada's core periphery structure at the provincial scale, a reference to political structures must be made. In this context, the implications for regional development of "being federal" must be recognized (Walker, 1980). The key point lies in the nature of the political role played by the regions. Because provinces choose their own leaders separately from elections which provide a mandate to federal authorities, these leaders have an independent position to develop and a power base with which to support it. Thus progress can only be made by discussion and compromise: the federal government can rarely ride roughshod over a provincial premier, with whom the majority do not agree, because other provincial leaders will always want to preserve the provincial right to retain an independent stance. The formal division of powers can vary greatly but in Canada both senior levels of government are involved in most important decisions relating to the economy, and meaningful unilateral action by one level is unthinkable.

In a federal country, such as Canada, regional grievances are invariably high on the political agenda. At the same time a federal structure means that to a significant degree there exists regional decentralization of political decision making, with all that this implies for local control and priority setting, job opportunities and investment expenditure. Admittedly, if provincial autonomy goes too far, provincial disparities may be exacerbated as the more powerful provinces exploit their greater bargaining strength while federal powers are correspondingly reduced. In general, however, provincial powers serve to provide a counter to the negative implications which arise from high levels of business concentration.

The core-periphery structure in Canada is not as clear-cut as it was 20 years ago, and could either strengthen or weaken in the future. Economically, Ontario is looking very strong at present, but there are powerful movements in the direction of greater political decentralization as evidenced by the recent debate on the Meech Lake Accord which would have weakened the power of the federal government on several fronts. Whatever the outcome of shifts in government power, business trends continue to move in the direction of concentration of ownership. In addition to political decentralization and business concentration, there are increasing calls for more autonomy at the local level. How these various forces are resolved will contribute greatly to the nature of Canada's core-periphery structure in the future.

5. TELECOMMUNICATIONS AND INTERNATIONAL ELECTRONIC INFORMATION SERVICES: AUSTRALIAN PERSPECTIVES

John V. Langdale

The impact of new information technologies on the global economy has been widely discussed. The argument generally presented is that we live in a global village and that the constraints of distance over communication have been largely eliminated. Furthermore, national institutions are becoming more heavily influenced by international forces. This global village argument, while attractive, is simplistic and even incorrect in a number of important respects. Changes taking place in the international information economy are quite complex. Rather than leading to a greater international orientation and uniformity of national institutions, the impacts of new information technologies are likely to produce a variety of outcomes—decentralization of some institutions and centralization of others.

There are a number of strands of geographical and related research that are useful in providing a framework for understanding changes in this area. Geographers have recognized the importance of the service sector in general (Daniels, 1985; Marshall, 1985). Some have focussed on the rapidly growing information sector as a key component of structural change in industrialized economies (Goddard, 1980; Kellerman, 1984). Hitherto, most geographical studies of the service and information sectors have examined particular regions or countries. An international focus is now necessary, given that an internationalization of production is emerging quite strongly in a number of information industries such as banking and finance (see also Friedmann and Wolff, 1982; Friedmann, 1986).

The purpose of this chapter is to examine Australia's international linkages in the business telecommunications and electronic information service

(EIS) areas. These linkages are considered in the framework of the emerging internationalization of production. Particular attention is focussed on the impact of internationalization of the rapidly growing banking and finance industry and its associated international electronic funds transfer (EFT) linkages. In addition, attention is directed towards Australia's linkages with the Asia-Pacific Region and the entry of Australian- and foreign-owned EIS transnational corporations (TNCs) in various countries in the region.

Services which rely on telecommunications and electronic information systems (EISs) are of particular interest because they are increasingly traded internationally. EISs are provided by service firms involved in the collection, processing and/or transmission of electronic information. While the distinction between EIS and non-electronic information service firms is difficult to make, it is clear that more firms are relying on electronic information systems.

EISs can be divided into two categories. Primary EISs include firms whose main function is the handling of electronic forms of information, such as banking, finance, computer services, electronic media (television and radio) and business and financial information. Other information service firms (accountants, advertisers and law), which still largely rely on non-electronic forms of information, are utilizing electronic information systems increasingly, especially the large firms. Secondary EISs are provided by firms primarily engaged in industries such as manufacturing, mining or other services (for example, health and transport). This category includes EISs which are provided on an intra-company basis. Indeed, firms in high-technology manufacturing areas, such as electronics, automobiles and aerospace, have quite sophisticated intra-firm electronic information systems linking production plants and offices throughout the world.

5.1 INTERNATIONALIZATION OF ELECTRONIC INFORMATION SERVICES

There has been rapid expansion in the volume of international trade in services. It has been estimated that world trade in the output of service industries, excluding foreign investment earnings, was over US$350,000 million in 1980 (US, 1984, p. 8). While this figure is small compared to an estimated US$1,650,000 million for merchandise trade, actual trade in services is likely to be much higher than the recorded figures. There are substantial conceptual problems in defining trade services, and there is a paucity of statistics in this area (Canada, Department of Communication, 1982, p. 6). Nevertheless, the rapid growth of trade in EISs is leading to an international telecommunications network becoming a key infrastructure in the international trading system.

Both primary and secondary EISs have become more important, although the relative importance of each is impossible to assess. While there are a few statistics available to document this growth, there is slightly more information on the growth and structure of primary EISs especially in the banking and finance area. Secondary EISs, being largely intertwined with the activities of TNCs in manufacturing, mining and other services, tend to be largely unreported. Consequently, this paper focusses on primary EISs.

The rapid growth of EISs internationally and within Australia, however, needs to be seen in the context of the internationalization of production in some manufacturing and service industries. Many globally integrated TNCs have upgraded their international information systems to link the various parts of their organizations. These developments have led to TNCs devoting more of their resources to secondary EISs in terms of intra-corporate financial, accounting and computer systems operations. The internationalization of production has also stimulated growth of primary EISs. Indeed, many TNCs in manufacturing and other service industries require high quality services from primary EIS firms which have had to improve their own international information systems in order to adequately service their clients' requirements (Langdale, 1987, p. 90).

Secondary EISs

Many, but by no means all, large TNCs in manufacturing, mining and services with extensive international operations have sophisticated international information systems and make extensive use of intra-corporate or secondary EISs. In general, high technology manufacturing firms in information (electronics, computer and telecommunications) equipment, automobiles and aerospace make extensive use of such systems. In the electronics industry, major manufacturers such as Motorola have linked Asian assembly and testing plants into their worldwide operations. (US, 1980, p. 630). Motorola's leased line network linking its North American, east and southeast Asian and European plants is an important component of its overall international manufacturing operations.

Some service industries (excluding EISs) are characterized by major international information flows. This characteristic is particularly true in the transport and travel and tourism industries. The former industry has introduced standardized procedures for documentation; the aim of such procedures is to facilitate the electronic transmission of bills of lading and other documentation. Most large international freight forwarders have extensive international corporate communications networks.

Aside from variations in the usage of secondary EISs by industry there appear to be variations by TNCs' country of origin. In particular, US-based TNCs have generally been faster in adopting sophisticated information technologies (especially for in-house use). This conclusion is supported by a study of the diffusion of international data communications technology among US- and European TNCs (Antonelli, 1985). Anecdotal evidence suggests that European and Japanese TNCs have improved their adoption rate of new information technologies in recent years.

Primary EISs

While there is a substantial diversity in the size, growth rate and nature of operations in different EISs most industrialized countries have shared in their growth. These services (and secondary EISs) tend to be geographically concentrated in major metropolitan areas (Daniels, 1985, p. 164) and in a manner which reflects an international urban hierarchy (Friedmann, 1986). In this latter regard, the geographical spread of TNCs' offices has been important in various countries. There has been a substantial increase in the volume of communications within these firms as well as between them and their clients.

It is difficult to recognize a common pattern in the internationalization of different EISs. International banks, for example, operate increasingly integrated global networks (Langdale, 1985a), a development facilitated by international EFT networks, deregulation and growth of offshore financial markets. In contrast, other EISs (law and accounting) operate in a far more decentralized manner largely because of the diversity of national regulations and business practices. Even in these cases, there has been growth in international operations; in part, because of the needs of TNC clients.

Electronic information service TNCs have also diversified the range of services provided. Banks, for example, are providing a wider range of financial services many of which are based on EFT networks. In accountancy, auditing work is becoming relatively less important for major firms as they move into consultancy and computer services. The reasons for this diversification relate to changes in technology and demand. Rapid adoption of electronic information systems by primary EIS firms and their clients is reducing barriers to entry. As EFT becomes more important, for example, the central position of banks in the financial system becomes less secure as computer service companies, merchant banks and other firms gain the technical capability to compete with them in particular areas. Many clients also want EIS firms to diversify their range of services.

5.2 AUSTRALIA'S INTERNATIONAL BUSINESS TELECOMMUNICATIONS LINKAGES

Australia's international telecommunications traffic is growing much faster than domestic traffic. From 1979-80 to 1988-89, for example, the growth rate for domestic telecommunications was 7.7 percent per annum for local telephone calls and 101.5 percent for international calls (Australian Telecommunication Commission, Annual Reports) and between 35.6 percent (telephone) and 16.1 percent for international (telex) traffic (Overseas Telecommunications Commission, Annual Reports). Unfortunately, traffic statistics on Australia's international business telecommunications are very limited and highly aggregated. The telephone service is used by both business and private subscribers, although the destination countries for the two groups tend to be different in a number of cases.

TABLE 5.1

AUSTRALIA'S OUTGOING INTERNATIONAL TELEPHONE TRAFFIC

	1965-66 Percent	1970-71 Percent	1975-76 Percent	1980-81 Percent
South-East Asia	1.36	5.42	5.50	5.82
East Asia	3.38	7.22	6.66	6.75
North America	26.30	23.05	15 58	17.08
New Zealand and Pacific Islands	32.40	28.07	25.85	21.54
Central and South America	0.10	0.21	0.39	0.54
TOTAL for Asia-Pacific Region	63.61	63.98	53.98	51.73
Europe	35.44	34.89	42.37	43.30
Rest of World	0.95	1.13	3.65	4.97
TOTAL[1]	**1,547,209**	**5,446,365**	**19,210,150**	**82,856,256**

Source: Unpublished OTC information.

[1] Total is expressed in terms of number of minutes.

The geographical composition of Australia's outgoing international traffic has changed substantially over time. The two major traffic destination regions are Asia-Pacific and Europe (Table 5.1). Despite the Asia-Pacific region's growing importance for Australia's international trade the region has declined in importance for telephone traffic and has remained relatively constant for telex traffic. Within the Asia-Pacific region there are substantial differences in the growth rate of international telecommunications traffic (Tables 5.1 and 5.2). Southeast Asia and, to a lesser extent, east Asia have shown strong growth, reflecting closer economic integration between Australia and the region (Langdale, 1984a, 1984b). Elsewhere, a strong growth in personal telephone traffic to Europe occurred in the 1970s, especially to countries which were the source of Australia's postwar immigration growth.

TABLE 5.2

AUSTRALIA'S OUTGOING INTERNATIONAL TELEX TRAFFIC

	1965-66 Percent	1970-71 Percent	1975-76 Percent	1980-81 Percent
South-East Asia	3.45	3.60	7.83	9.46
East Asia	25. 88	15.04	16.66	16.91
North America	23.90	24.47	23.88	22.56
New Zealand and Pacific Islands	9.90	12.38	15.86	12.90
Central and South America	0.26	0.17	0.31	0.30
TOTAL for Asia-Pacific Region	63.39	55.66	64.54	62.14
Europe	34.52	42.12	31.22	30.01
Rest of World	2.09	2.22	4.24	7.85
TOTAL[1]	**764,832**	**3,582,456**	**9,564,983**	**20,030,073**

Source: Unpublished OTC information.

[1] Total is expressed in terms of number of minutes.

Surprisingly, North America's relative importance has declined despite the close economic ties and the importance of information-intensive US-based TNCs operations in Australia. It is likely, however, that a substantial amount of business traffic has been diverted to leased lines: these are lines leased for the use of an organization and are used only for intra-corporate purposes.

International leased circuits have grown substantially from 217 in 1976 to 527 in 1985 (Overseas Telecommunications Commission, Annual Reports). There are two types of leased circuits: slow-speed telegraph and medium-speed voice/voice data. While there are fewer voice-grade leases their growth rate has been higher. (In 1977 there were 203 telegraph and 37 voice-grade leases; by 1985 there were 361 and 89 leases respectively). The geographical distribution of both telegraph and voice-grade circuits is dominated by countries in the Asia-Pacific region. In 1983, for example, Hong Kong accounted for 24.4 percent of telegraph leases. While Hong Kong is of increasing importance to Australia as an international banking and finance centre, the concentration of leased circuits between the two countries reflects Hong Kong's favourable international telecommunications rates. Many TNCs use a Sydney/Melbourne to Hong Kong leased circuit to connect with their global telecommunications networks.

The US has the largest number of medium speed voice-grade links with Australia, accounting for 44.9 percent of such links in 1983, primarily because of the greater technical sophistication of international information systems of US-based TNCs, many of which are concentrated in high-technology industries. Since there are relatively few leased circuits from Australia to Japan a number of Japanese firms use Hong Kong as their Asian telecommunications hub. Leases to other countries in the Asia-Pacific region, such as New Zealand, Fiji and Papua New Guinea, reflect the operations of Australian- and foreign-owned TNCs such as banks, finance, insurance and airline companies, as well as Australian government departments' activities in the region.

A small but rapidly growing component of international telecommunications is the use of international data bases. The volume of traffic on this service (Data Access) grew from 3.774 million to 4.787 million connect minutes in 1984-85, a growth of 26.8 percent (Overseas Telecommunications Commission, Annual Report). Australian users have a variety of means of accessing these data bases: many use the Data Access service offered by OTC; others access data bases using facilities provided by computer service bureaus (for example, I. P. Sharp and GEISCO), transnational banks, or business and financial information providers.

The geographical distribution of usage on the Data Access service is heavily dominated by the US which accounted for about 95 percent of incoming and outgoing links in 1983. This reflects that country's dominance of the international computer services market (US International Trade Administration, 1984). In addition usage is predominantly from Australia to the US. The small volume of incoming traffic largely involves the overseas head offices of US-based TNCs accessing the Australian subsidiaries' computer data bases.

Sydney, and to a lesser extent Melbourne, dominate Australia's international telecommunications. For incoming international telephone traffic in June 1983, Sydney (02 area code) attracted 38.3 percent and Melbourne

(03 area code) 23.5 percent of the national total. These two cities together with Perth (8.2 percent), Brisbane (6.7 percent) and Adelaide (4.5 percent) attracted 81.2 percent cent of the total incoming telephone traffic. Sydney's dominance is higher for business-oriented international telecommunications services. For outgoing international telex traffic in March 1981 over 50 percent of traffic originated from the Central Coastal Region of New South Wales (which includes Sydney, Newcastle and Wollongong) compared with 30 percent for Melbourne and 14 percent for all other capital cities (Langdale, 1982a, p.79). The dominance of Sydney is even greater for international leased line services. Of the total number of leases in March 1981 with an identifiable geographical origin/destination, New South Wales (principally Sydney) had 59.5 percent, Victoria (principally Melbourne) 19.8 percent and Canberra 17.2 percent. Many companies and government departments link their international and national leased line networks. Thus branch offices and plants in peripheral states have international telecommunications access similar to the Sydney, Melbourne and Canberra offices.

5.3 AUSTRALIA'S INTERNATIONAL TRADE IN EIS

Australia's international trade in EISs is examined in the context of trends in the internationalization of EISs and the changing patterns of international telecommunications outlined in earlier sections. These trends are leading to Australia's becoming more tightly linked into the international information economy.

The internationalization of manufacturing and other industries is creating pressures for the internationalization of primary EISs. Australian-owned primary EIS firms are responding to these pressures by establishing overseas branches, especially in the Asia-Pacific region, but also in major international business and financial cities such as New York and London. They are also linking these branch networks with international telecommunications. In the case of larger firms, they have leased extensive networks. Moreover, competition from foreign-owned electronic information service TNCs with branch offices in Australia is forcing Australian EIS firms to improve the quality of their international services.

There are significant interdependencies among primary EIS firms, and changes in the financial environment can have far reaching effects. The floating of the Australian dollar and other deregulatory financial policies of the Australian government, for example, have resulted in a rapid growth in activity for business and financial information providers. Rapid fluctuations in the value of the Australian dollar have led to a proliferation of terminals accessing services provided by such firms as Reuters and Telerate. The majority of these terminals are likely to be located in the decision-making centres of Sydney and Melbourne. Expansion of electronic information services provided by Reuters and Telerate also has significant impacts on the financial system. Indeed, the future role of stock exchanges is in question given the technical capability of Reuters or Telerate to provide electronic domestic and/or international trading in securities.

Australia's international trade in EISs, as is the case with telecommunications traffic, is focussed on Sydney and Melbourne with lesser concentrations in other capital cities. Sydney and Melbourne's dominance in Australia's international trade in EISs can be linked to their key role in both primary and secondary EISs. To provide more insights into the dynamics of change specific primary EISs, notably in the financial and legal industries, are now considered.

BANKING AND FINANCE

Australia's international linkages in banking and financial services have increased substantially in the past 10 years as a result of international banking and finance firms establishing operations in Australia and linking these activities to international financial markets, and Australian firms expanding overseas, especially in the Asia-Pacific region. With respect to the former type of development, Australia clearly represents an important market for major transnational banks (especially from the US, UK and Japan) given the large number of TNCs in manufacturing, mining and other industries operating in Australia which often use these banks on a worldwide basis. Foreign banking and finance firms have rapidly expanded their operations in Australia, with merchant banks and finance companies growing during the 1960s and 1970s and with the government's decision in 1985 to allow 17 foreign banks to commence operations, joining the two foreign banks already in operation. Sydney has 10 of the head offices, Melbourne has five (two banks have a joint Sydney-Melbourne head office) and Perth and Adelaide each have one.

The international telecommunications linkages of foreign banks and financial institutions are quite extensive, particularly those of large US financial institutions (Citicorp, Bank of America and Bankers Trust). They are heavy users of international leased networks, but also make substantial use of international public switched services (telephone and telex) and SWIFT, the co-operatively owned interbank network (Langdale, 1985a).

With respect to the rapid internationalization of Australian banks in recent years, the largest number of their overseas branch and representative offices are concentrated in the southwest Pacific region comprising New Zealand, Papua New Guinea and Fiji, and they are largely retail-type operations (Hirst, Taylor and Thrift, 1982). The main focus of their recent overseas expansion has been in the major international financial centres and in the east and southeast Asian region (Langdale, 1985b, p. 78). ANZ's takeover of Grindlay's Bank in 1984 gave it an edge over other Australian banks in its internationalization strategy (Hirst and Taylor, 1985, p. 294). However, Westpac has recently acquired a number of US and UK financial institutions and diversified its international activities. The Australian banks have also expanded their usage of international telecommunications services in order to provide a range of electronic services for their corporate customers similar to those provided by large foreign banks.

SECURITIES INDUSTRY

Internationalization of the international securities industry combined with deregulation of the Australian stockbrokerage industry is leading to major changes. International securities firms are expanding their Australian operations. Similarly, several Australian firms are following expansionary strategies and are widening their geographical representation in Australia and in the major international financial centres of London and New York. Some have merged with other Australian stockbrokers. The National Australia Bank has taken a 100 percent holding in the large Melbourne-based company A. C. Goode. In other cases, linkages have been established with foreign firms, especially UK-based firms. The large Australian firm, Potter Partners, for example, has sold a 50 percent share to the recently formed UK-based international securities group Mercury International, comprising member companies Warburg, Rowe and Pitman, Arkroyd and Smithers, and Mullins (Kennedy, 1986). Such foreign linkages give Australian firms a much stronger presence in major world capital markets (Powditch, 1985).

A limited expansion of Japanese securities firms in Australia is likely, given the growing internationalization of such firms as Nomura, Daiwa and Nikko. By contrast, US securities firms in Australia are less likely to expand given that several are rationalizing their international operations following substantial losses. For their part, Australian stockbrokers have been active in expanding international operations although these are confined to major international financial centres, principally London with 22 firms in late 1985 as against seven in New York (Korporaal, 1985). There is also some representation in the Asia-Pacific centres, notably Hong Kong, but also Singapore and New Zealand. A less common overseas expansion strategy is that followed by the Australian firm Roach Tilley Grice, which is 40 percent owned by the diversified Australian financial and manufacturing conglomerate Elders IXL. It has taken a 50 percent ownership of a small UK broker with the aim of expanding rapidly in European markets. This move needs to be seen with reference to Elders' overall expansion as an international and domestic financial service conglomerate (see Chapter 12 of this volume).

FUTURES INDUSTRY

The Sydney Futures Exchange in the mid- to late 1980s attempted to establish links with international futures exchanges to form the basis of a 24- hour global market. For example, the Exchange was linked with the London International Financial Futures Exchange (LIFFE) so that dealers could trade in Eurodollar deposits and US Treasury bonds. However, these linkages have not proved to be successful and most have now been abandoned.

In common with other areas in international banking and finance, there is considerable rivalry among countries in the Asia-Pacific region. The main rivals to Sydney are Singapore (SIMEX), Hong Kong and Tokyo. At present the other Asia-Pacific futures exchanges are attempting to attract offshore Japanese business. Singapore introduced a contract based on the Nikkei average of 225 Japanese equities with the aim of tapping Japanese financial institutions wishing to trade in an offshore location (Duthie, 1986).

It is difficult to forecast which centre will be the winner in this rivalry. While Sydney has a two- to three- hour advantage in opening earlier than Singapore and Tokyo, it has an equivalent disadvantage with respect to London's opening. The Sydney Futures Exchange has the advantage of a large domestic market as compared with Singapore, but the latter has a large offshore international banking presence. While there has been considerable activity in introducing new contracts for the various exchanges, it is not clear that there is, as yet, much demand for 24-hour trading of future contracts.

Certainly, Singapore's SIMEX link with the Chicago Mercantile Exchange has not lived up to expectations although it has not been a failure.

Other EISs

In common with the banking and finance industry there has been a general trend towards the internationalization of other EISs by both foreign- and Australian-owned EIS TNCs. In fact, recent expansion of electronic information service TNCs in Australia of business and financial information providers, accounting, computer service and legal firms has been associated with developments in banking and finance. Most of these firms have their Australian head offices in Sydney or Melbourne but are also expanding nationally, especially in association with considerable resource extraction and processing activity in Queensland and Western Australia.

LEGAL SERVICES

Legal services illustrates a number of relevant trends. It is not strictly in the EIS category since information is generally handled via personal contact or in written form. However, electronic information is becoming more important, especially in the larger legal firms. A few of the larger Australian firms are represented in London, New York and Singapore (Davies, 1986) while a number of foreign-based legal firms (Coudert Bros. and Baker and McKenzie from the US) have established offices in Australia. Initially these offices were limited to international and US law (as is the case in Singapore). However, in late 1985 Coudert Bros. was allowed by the New South Wales Law Society to practice local law in that state.

It is unlikely that international telecommunications linkages in the legal services area are very large at present. Some firms (e.g., Coudert Bros.) operate a single global partnership (like Arthur Andersen in the international accounting area) and are likely to generate considerable international telecommunications traffic. In contrast, other firms (Baker and McKenzie) operate as a number of largely separate national partnerships under a common name. However, further internationalization of client firms (especially with the entry of foreign banks in Australia) is likely to alter this situation. Large law firms transfer a substantial amount of electronic information between their national and international offices; they also access foreign and local data bases and rely more heavily on telecommunications in dealing with clients.

5.4 DEREGULATION OF INTERNATIONAL TELECOMMUNICATIONS AND TRADE IN EISs

A number of changes in the international regulatory environment in telecommunications and EISs are having a significant impact on Australia's international linkages. In this regard, deregulation of international telecommunications and EISs is related to the thrust by the US and other major industrialized countries to liberalize trade in services. Deregulation of the international banking and finance market is but one example of many services in which these countries see their firms as having greater access to international markets in a deregulated environment. The US is particularly interested in liberalizing trade in EISs since a number of US-based TNCs have a strong competitive advantage (US National Telecommunications and Information Administration, 1983; US, 1984).

A closely associated issue is the introduction of competition in international telecommunications (Langdale, 1982b). Large international telecommunications users (chiefly TNCs) and governments of major industrialized countries (US, UK and Japan) are pressing for the introduction of competition. A competitive environment in international telecommunications is likely to enhance the status of the UK, US and Japan as international telecommunications hub countries as well as attracting additional EISs (Langdale, 1985b, pp. 12-17). Competition is likely to be fiercest on the heavily trafficked north Atlantic route between the US and western Europe and to a lesser extent on the north Pacific route between the US and Japan. In a competitive environment, telecommunications rates for large users (principally TNCs) on these routes would be substantially reduced.

It is not possible to discuss details of proposed competitive developments in this paper. Suffice to say a number of private US satellites have been given permission in principle by the US government to offer private lines to users (chiefly TNCs and media companies). Users would not be allowed to connect these lines to the public switched network. The private satellite companies are primarily interested in the US to western Europe market although other proposals will connect the US to Latin America and the

Pacific region to Japan in particular. In addition, Cable and Wireless, in conjunction with a number of other companies, is planning a privately-owned worldwide fibre optic network linking the UK with the US and the US to Japan (Crisp, 1986).

Australia, being distant from the heavily trafficked routes, is unlikely to attract a substantial level of competition. Australia's international users of telecommunications (primary and secondary EISs as well as private users) would be relatively disadvantaged in such an environment. Similarly, smaller firms are likely to be disadvantaged in a competitive environment which would particularly hinder the efforts of small Australian EIS firms to internationalize their operations. They would also be faced with relatively higher telecommunications rates to the region of greatest interest in building up new trading linkages namely, countries in southeast and east Asia and the Pacific Islands.

5.5 CONCLUSION

Global changes in demand for international information transfer on the part of large organizations combined with rapid technological change in information equipment and shifts towards deregulation of trade in services have had a major impact on the size and nature of Australia's international (and national) telecommunications linkages. The volume of Australia's international telecommunications is likely to rise even more rapidly in the future, primarily in the business telecommunications area. OTC estimates that it carried in 1986 approximately 1,000 giga (109) characters of international information. By 1991 it expects this to exceed 5,000 giga characters and by 1996 over 40,000 (Robins, 1986).

It would be expected that the geographical distribution of Australia's international telecommunications traffic, especially in the business telecommunications area, will show a greater concentration on the Asia-Pacific region in the future despite the mixed trends with respect to the region in the past. Future traffic growth is likely to come from both the primary and secondary EIS areas as Australian-based firms expand in the region. Linkages between Australia and the region, however, will be strongly influenced by global trends in the internationalization of EISs. We have already seen that electronic information service TNCs from the US and to a lesser extent western Europe have expanded in the Asia-Pacific region. A question of major importance is: To what extent are Japanese primary and secondary electronic information service TNCs likely to be successful in internationalizing their operations? There are already strong indications that the Japanese government has targeted these industries for rapid growth. Deregulation of the Japanese financial and telecommunications industries should be seen in the context of their goal of shifting into high value-added EISs.

These rapid changes in the international information economy are altering Australia's and other countries' relative cost distances to the rest of the world. Many have argued that we live in a world in which the costs of distance are shrinking partly as a result of the adoption of new information technologies. In practice, while some of these cost distances are likely to decline very rapidly for some regions (the north Atlantic and north Pacific) and for large users, in peripheral regions and for small users they may increase.

These changes pose significant questions for geographers. The emerging international information economy is leading to a major restructuring of international and national cost distances. Current and projected changes in such factors as information technology, the nature of demand for international information transfer and government deregulatory policies indicate that the pace of restructuring in the nature of cost distances is likely to increase in the future.

DIMENSIONS OF EMPLOYMENT AND SOCIAL CHANGES

6. INTERNATIONAL INFLUENCES ON REGIONAL UNEMPLOYMENT PATTERNS IN CANADA DURING THE 1981-84 RECESSION

Glen Norcliffe and Donna Smith Featherstone

This chapter is concerned with foreign influences on patterns of regional unemployment in Canada during the recent recession. It will be shown that in an economy as open as Canada's, the particular way in which a region responded to the global recession of the early 1980s was influenced by factors both internal and external to each region. The degree to which a regional economy fared better or worse than the Canadian average during the recession was directly related, in many cases, to a region's employment specializations, and to foreign trade performance in those specializations. This is not to dismiss local factors which were of varying importance. But it is to assert that a combination of factors including high levels of penetration by foreign capital, a huge degree of integration (in certain sectors) into the US economy, a penchant for exporting staples and intermediate products, and an associated weakness in exporting end products, all conspired to increase the vulnerability of Canadian regions to cyclical events originating elsewhere.

6.1 THE 1981 RECESSION

The economic slump that marked the beginning of the 1980s might better be labelled a depression than a recession, for to some degree it had an impact on all corners of the globe. It was seen by some observers as the end of a long (Kondratieff) wave that began with the Great Depression in 1931, peaked in the prolonged postwar boom, and ended with high inflation followed by soaring unemployment in the early 1980s. The impact and timing of the recession varied from country to country. In the UK, for instance, the rigid monetarist policies of the Thatcher government which used rising unemployment as a policy instrument to combat inflation, brought about recession as early as 1978. In Canada, the picture was somewhat different as Figure 6.1 shows. The period from 1975-1981 was one of chronic unemployment, with the national rate mostly in the 7-8 percent range. Eventually this level came to be accepted as a "norm" which reflected growing frictional and structural unemployment in a dispersed but technologically sophisticated economy (Donner, 1986). Any complacency about Canada's economic performance evaporated in late 1981 as national unemployment skyrocketed to almost 13 percent in late 1982. As will be demonstrated, the impact of the recession varied greatly from region to region.

For analytical purposes it is useful to divide the recession into two phases, a downswing and a recovery. The downswing was much more uniform in its timing than the recovery (Norcliffe, 1987). Using the method of turning points (Bassett and Haggett, 1971; Lever, 1980), April 1981 was identified as the month when national unemployment turned upwards. Fully 39 out of Canada's 53 Labour Force Economic Regions (74 percent of the total) went into recession within four months either side of this national turning point. Among the 11 regions that went into recession more than four months before the nation, five were in the Atlantic provinces, and three were regions in Ontario that had a substantial manufacturing base. Only three regions withstood the tide of the recession later than August 1981: Edmonton still basked in the last rays of the oil boom; the Hull region continued to benefit from the expansion of federal government activities on the east side of the Ottawa River; and the primary metal industry in Quebec's Lac St. Jean region expanded its employment between early 1981 and early 1982. Overall, however, the great majority of Canada's regions marched to the tune of national recession.

The recovery was a very different story, with timing of regional upswings varying greatly from region to region. The standard deviation for the timing of regional recovery was 11.6 months or three times that of the downswing. As will be shown later, one of the key factors governing recovery was the export performance of a region's main industries. In western Canada, the only region to recover early was the Prince George region in which the very large Quintette coal deposit is located. Supported by Japanese investment and purchase contracts, this project and associated infrastructure developments had created many new jobs by mid-1984. Elsewhere in western Canada, the recovery came later, and was delayed by a prolonged decline in

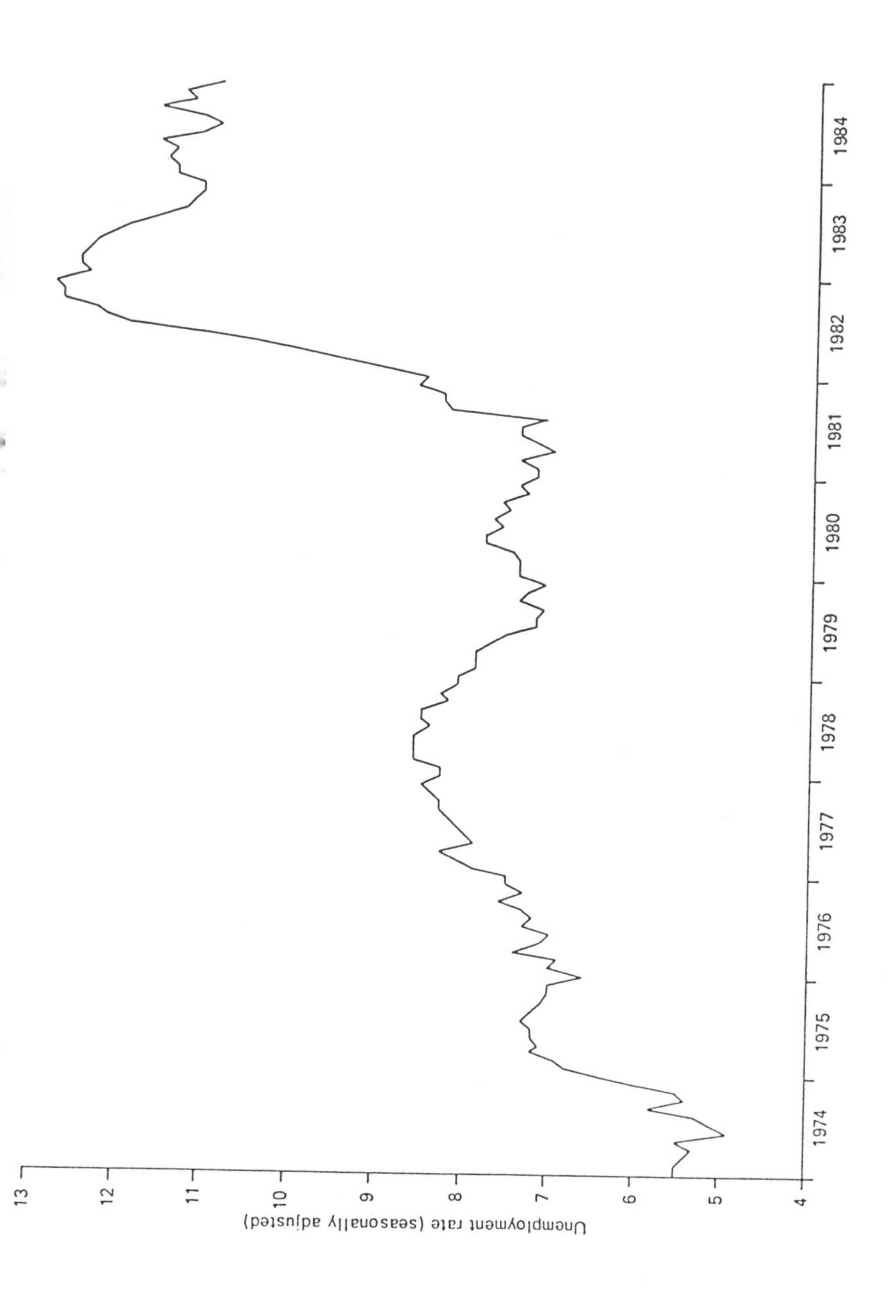

Figure 6.1 Seasonally adjusted unemployment in Canada 1974-1984

the international price of oil (up to late 1986) and by stagnation in agricultural and forestry exports in 1984.

Central Canada dictated the pace of national recovery, and indeed was the only part of the country to show a uniform pattern of recovery. All of the Labour Force Regions in Ontario and southwest Quebec began their recovery within four months of the Canadian upturn (which may be somewhat tautological since this region accounts for nearly two thirds of Canadian GDP). Nevertheless, there was a significant difference between central Canada's fairly uniform recovery and the heterogeneous recovery pattern found in other regions.

The timing of recovery in eastern Canada was highly variable, but generally late. Newfoundland and Labrador languished in recession longer than any other province; there was a prolonged slump in the fishing industry, while at the time the offshore Hibernian oil discoveries amounted to little more than gold at the end of the rainbow. Cape Breton's economy also remained in deep recession, unable to drum up significant orders for steel from the ailing Sysco plant. Montreal, meanwhile, by succeeding in establishing itself as a 12-month deep-sea port, reduced activity levels in the major traditional east coast winter deep-sea ports of Saint John and Halifax. In addition, the apparent early recovery of two mining regions in northern Quebec was in practice the result of the outmigration of mine employees who had been laid off permanently (Bradbury and St. Martin, 1983); this lowered the unemployment rate, but there was no real economic recovery. In short, the economy of eastern Canada lay in partial paralysis long after central Canada had begun to swing back.

Besides timing, the other main regional aspect of the recession was how severely it affected a region. Such cyclical sensitivity can be measured in several ways. Bear in mind that labour markets show rates of labour turnover far in excess of absolute employment levels (Martin, 1984). Thus there is considerable "circulation" in the labour force with additions composed of entrants and re-entrants, and withdrawals composed of retired and discouraged workers. A layoff may result from voluntary leaving or involuntary redundancy, and new employment may be due to a new hire or a rehire of someone previously employed. Unemployment is defined as the balance of this circulation. Unfortunately, Statistics Canada data do not begin to allow approximation of these components of employment circulation at the regional level, which leaves three other methods of estimating cyclical sensitivity. The first is a regional employment elasticity coefficient of the form

$$\eta_r = \frac{\delta u_{r,t}}{\delta u_{n,t}}$$

where δu is the rate of change in unemployment in region r (and the nation, n) over time period t. ηr is > 1 when a region's economy is cyclically sensitive, whereas the reverse is the case when the elasticity coefficient is less than unity. The time series available in this case does not permit reliable estimation of this elasticity coefficient, though it has interesting possibilities for longer time series.

The second measure of cyclical sensitivity is the Brechling coefficient (Brechling 1967; King and Clark, 1978), which is used in Norcliffe (1988) to describe the present case. But the simplest and most direct measure is a performance index of the form

$$P_r = \frac{U_r = (t+1) - U_r (t)}{U_n (t+1) - U_n (t)}$$

Figures 6.2, 6.3 and Table 6.1 plot the performance indices for the downswing and recovery, respectively. Three main groups of regions performed poorly during the downswing: British Columbia was hit by a severe recession in the forest product industry; the mining and forest product industries of the Shield in Ontario and Quebec and in New Brunswick suffered a serious setback; and the three industrial regions on the North Shore of Lake Erie—Windsor, London, and Hamilton-Niagara — experienced soaring unemployment, affected particularly by layoffs in the primary metal and vehicle industries and in those linked to it. The Prairies, in contrast performed comparatively well, the worst consequences of the recession being staved off by good wheat sales and the diffusion of "new" crops such as canola, soy and triticale. Newfoundland, too, rode the storm waves of the slump quite well, with the fishing industry still enjoying some benefits attributable to the declaration of a 100-mile fishing limit. Finally, the fairly diversified labour force regions that stretch from Kitchener-Waterloo along the north shore of Lake Ontario to Montreal also performed better than average during the downswing.

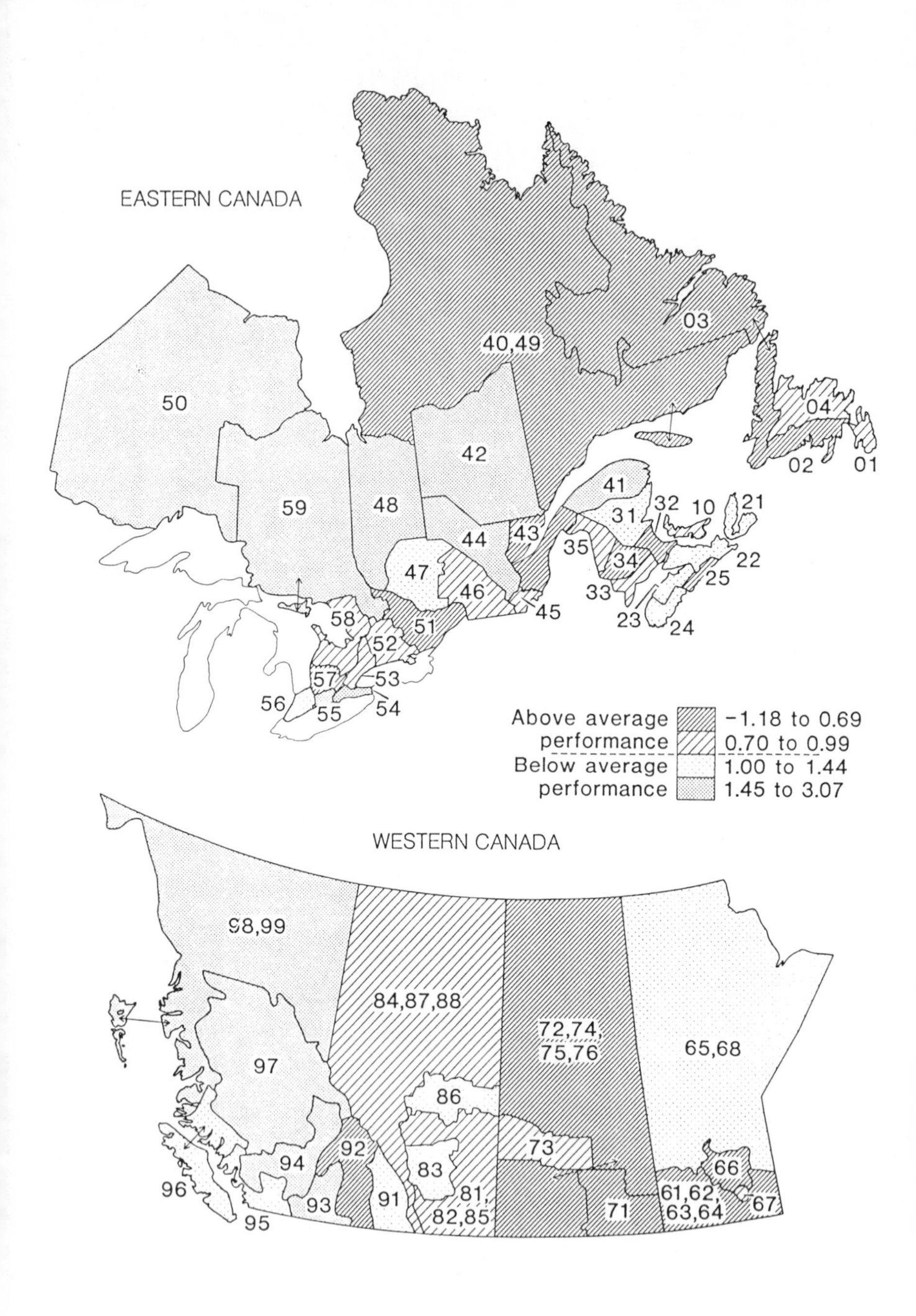

Figure 6.2 Unemployment performance in Labour Force Economic Regions during the downswing (August 1981 - December 1982)

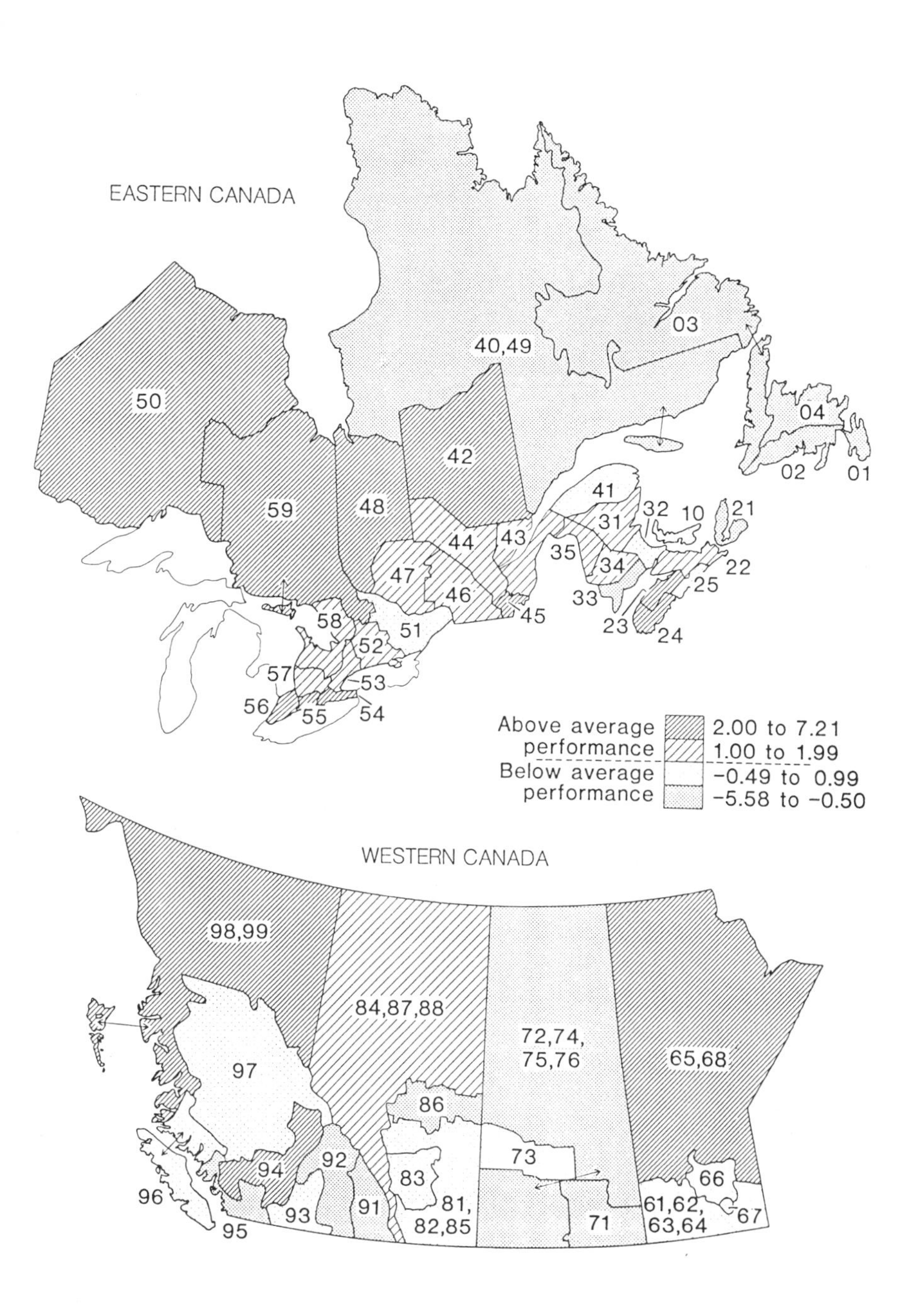

Figure 6.3 Unemployment performance in Labour Force Economic Regions during the recovery (December 1982 - December 1984).

Nationally, the economic recovery began in December 1982. However, this recovery could hardly be described as an economic boom, for two reasons. First, it affected only certain regions. Those regions with a negative score in Figure 6.3 (one third of the total) had higher unemployment levels in December 1984 than in December 1982; indeed most of these regions had begun to recover only towards the end of this period. The standard deviation of the performance index in the recovery phase was 2.12, compared to 0.70 during the downswing, again indicating the great variability of regional performance after December 1982. And second, the recovery was at best partial so that even by December 1986 the seasonally adjusted national unemployment rate had dropped to only 9.4 percent. (Of necessity, this regional analysis ends in December 1984 when a substantially revised system of Labour Force Economic Regions was adopted by Statistics Canada).

Recovery was largely confined to central Canada, and was most marked in regions that fared worst in the downswing, thereby attesting to the cyclicality of their economic performance. The three regions on the north shore of Lake Erie that experienced big increases in unemployment during the slump regained most of the ground they had lost due mainly to a boom in the vehicle industry. The mining and forestry regions on the Shield also recorded a sharp drop in unemployment attributable in part to higher activity levels, but also helped by local declines in the size of the total labour force due to out-migration and lower labour force participation rates.

Elsewhere in Canada, good recoveries were achieved in only a few localities. Coal developments in the west (Gregg River in northern Alberta, and Quintette in northern British Columbia) triggered economic growth both locally and in Prince Rupert where major investments were made in port infrastructure. Two regions in the southern part of Nova Scotia posted a strong recovery, probably due to a number of small but favourable developments including a flourishing tourist industry which was assisted by a decline in the value of the Canadian dollar.

Other regions in the east and the west remained in the grip of the recession into 1983 and some even into 1984. Continued low export demand for forest products kept much of BC in recession. The picture in the Prairies was particularly gloomy with cutbacks in oil and gas exploration and depressed prices for agricultural products due to overproduction and subsidies in the EEC and the US. Continued problems in offshore oil exploration, the east coast fisheries and the forest product industries kept several of the Atlantic regions (Newfoundland in particular) in recession long after central Canada had achieved a significant recovery.

To summarize, regional performance during the 1981-84 recession was asymmetrical. The downswing was relatively uniform in its impact: it affected most regions in mid-1981, and although the force of that impact varied somewhat, few regions marched much out of step. However, in the 20 months of national recovery for which consistent data are available, the record is highly variable with regions marching in several different directions. The timing of recovery was notably variable. And to a large degree central Canada basked in economic sunshine while the rest of the country continued to suffer economic squalls. The next step in the analysis is to try to throw some light on this issue of regional variability.

6.2 THE ROLE OF INDUSTRIAL STRUCTURE

It may seem intuitively plausible that the mix of industries in a region would have an influence on the cyclical sensitivity of employment levels, for the simple reason that industries respond differently during a recession. Unfortunately, those studies which have sought to identify this relationship have found that the role of industrial structure is "virtually irrelevant" (Cheshire, 1973), "very small" (Armstrong and Taylor, 1978), or "minor" (Taylor and Bradley, 1983), and that unemployment performance "cannot be reduced to a sectoral interpretation" (Marchand, 1986). Thus it might seem to be flying in the face of conventional wisdom to seek an explanation based on differences in the regional mix of industries. However, Townsend's (1983) study of unemployment in the UK during the 1979-81 recession found industry mix to be an important variable. Accordingly, Norcliffe (1988) has examined the role of industrial structure in Canadian regions in the 1981-84 period. The results are summarized below, since they are a necessary step to examining international influences in the final section.

The first step in the analysis was to separate cyclical and structural unemployment. This was achieved using an additive Brechling model of the form:

$$U_r = B_{r,o} + B_{r,1}\, U_n + E_r$$

where U_r is the level of unemployment in region r

$B_{r,o}$ and $B_{r,1}$ are regional regression coefficients estimated using OLS procedures and U_n is the national equilibrium (structural) unemployment rate, estimated to be 6 percent.

$B_{r,1}$ is an estimate of regional cyclical sensitivity. A value of 1.0 indicates cyclicality that matches the nation; values above unity indicate high cyclical sensitivity, while values below unity indicate less than average sensitivity. $B_{r,0}$ is a measure of structural unemployment: it indicates how much, on average, a region's "equilibrium" unemployment rate stood above or below the national figure. Broadly speaking, the major regions can be classified in a 2 x 2 table as in Figure 4. Labour Force Economic Regions of particular interest include Hamilton-Niagara (which is cyclically sensitive), Victoria (which is the least cyclical region in British Columbia thanks to provincial government employment), and Calgary and Edmonton, both of which are highly cyclical (details are reported in Norcliffe, 1988).

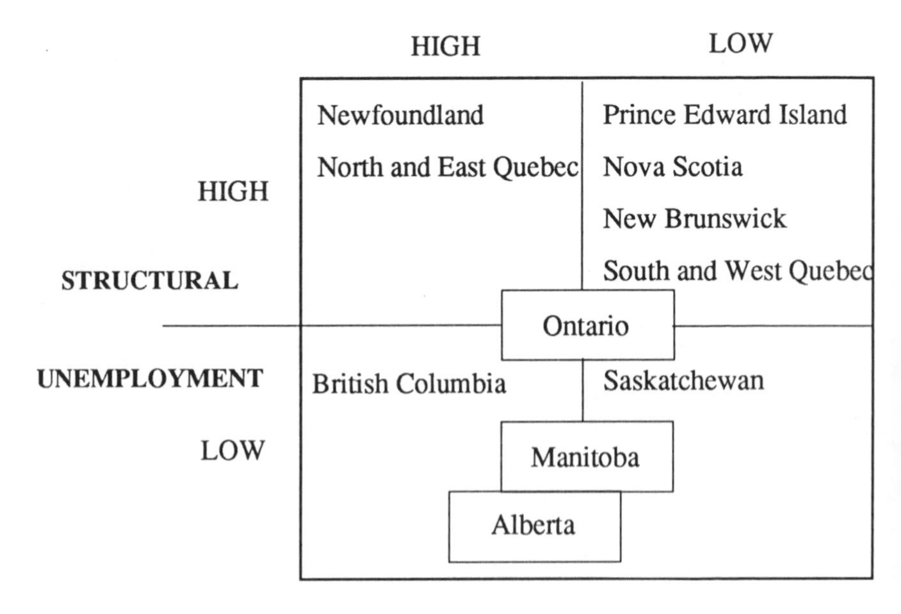

Figure 6.4 Major regions classified by type of unemployment.

How important is industry mix to this cyclicality evident in so many of Canada's economic regions? There are, in practice, two elements of cyclicality to be considered. As already demonstrated, unemployment change (Δ U) is cyclical. But so is labour force participation, defined as the change in the size of the labour force (Δ LF). Put together, we obtain more formally

$$\Delta E_{r,t} = \Delta LF_{r,t} - \Delta U_{r,t}$$

hence changes in total employment capture both cyclical effects. With this formulation, the influence of industry mix can be readily isolated using standard shift-share methods so that

$$\Delta E_{r,t} = \Delta S_{r,t} + \Delta C_{r,t}$$

where S and C represent the structural and competitive components of employment change.

The structural components were then inserted as an independent variable in a simple regression model predicting employment change for each of the three years 1981-82, 1982-83, and 1983-84 (see Norcliffe (1988) for details). The results indicate that industry mix accounted for 23, 27, and 16 percent of the variance in the three years, respectively, in all cases a significant proportion.

This result is supported by a detailed empirical examination of the consequences of short-run changes in specific industries on the regions in which they are located. A few of the most vivid examples will be presented since they provide a connection to the final section of this paper. The examples relate to employment changes in 12 major activity categories over the three annual periods, 1981-82, 1982-83, and 1983-84, based on the fourth quarter of each year (see Table 6.2).

TABLE 6.1
SEASONALLY ADJUSTED UNEMPLOYMENT RATES, PERFORMANCE INDEXES, AND LEADS AND LAGS, BY REGION

	UNEMPLOYMENT RATE			PERFORMANCE INDEX		PHASING	
	AUG.	DEC.	DEC.				
Region	1981	1982	1984	Downswing	Recovery	Downswing	Recovery
Newfoundland							
01	12.6	16.6	18.7	0.77	-1.11	14	57
02	13.0	16.9	27.5	0.68	-5.58	16	57
03	15.9	18.8	20.8	0.51	-1.05	17	53
04	16.	21.4	23.7	0.88	-1.21	10	59
Prince Edward Island							
11	8.9	13.2	13.5	0.75	-0.16	20	28
Nova Scotia							
21	14.2	20.7	21.9	1.14	-0.63	9	43
22	10.3	16.6	14.2	1.11	1.26	5	40
23	8.1	15.7	11.9	1.33	2.00	18	50
24	8.3	14.4	8.6	1.07	3.05	18	37
15	7.9	11.1	11.6	0.56	-0.26	13	59
New Brunswick							
31	12.3	18.5	16.2	1.09	1.21	14	40
32	12.5	13.9	14.3	0.25	-0.21	9	39
33	7.0	12.0	15.6	0.88	-1.89	20	60
34	11.1	11.8	9.8	0.12	1.05	12	40
35	10.5	15.9	12.5	0.95	1.79	10	28
Quebec							
40	20.5	13.8	20.8	-1.18	-3.68	18	22
41	12.6	25.5	23.7	2.26	0.95	8	40
42	11.2	19.8	15.1	1.51	2.47	26	41
43	9.2	12.3	9.3	0.54	1.58	20	31
44	10.0	18.4	15.8	1.47	1.37	19	38
45	11.6	16.2	7.6	0.81	4.53	19	34
46	8.9	14.0	11.5	0.89	1.32	16	34
47	11.1	17.7	14.0	1.16	1.95	26	36
48	8.6	23.7	16.4	2.65	3.84	20	30

Ontario							
50	4.6	14.5	9.1	1.74	2.84	20	38
51	6.6	9.6	9.5	0.53	0.05	17	40
52	9.4	23.4	10.1	0.70	1.74	9	39
53	4.5	9.7	6.3	0.91	1.79	12	40
54	7.8	18.2	10.3	1.82	4.16	11	36
55	5.3	13.6	9.7	1.46	2.05	20	37
56	7.6	15.5	9.3	1.39	3.26	20	37
57	5.8	9.4	7.1	0.63	1.21	11	37
58	6.6	11.3	8.9	0.82	1.26	19	40
59	6.4	19.4	14.7	2.28	2.47	20	32
Manitoba							
61	4.5	7.9	6.5	0.60	0.74	18	50
65	5.4	12.8	6.3	1.30	3.42	13	35
66	6.3	8.6	8.3	0.40	0.16	18	58
67	6.5	12.0	10.4	0.96	0.84	12	36
Saskatchewan							
71	4.1	6.2	8.1	0.37	-1.00	17	40
72	3.6	6.9	7.9	0.58	-0.53	19	51
73	5.0	9.5	8.9	0.79	0.32	9	57
Alberta							
81	2.3	6.4	6.9	0.72	-0.26	20	43
83	3.0	10.9	10.2	1.39	0.37	18	57
84	3.7	9.2	6.6	0.96	1.37	18	49
86	4.0	10.1	12.8	1.07	-1.42	22	53
British Columbia							
91	6.2	14.0	15.8	1.37	-0.95	12	57
92	14.5	15.0	16.6	0.09	-0.84	14	51
93	6.6	16.5	15.8	1.74	0.37	17	38
94	10.0	27.5	18.8	3.07	4.58	16	36
95	4.9	13.1	14.5	1.44	-0.74	13	54
96	8.1	16.3	14.7	1.44	0.84	11	50
97	6.3	17.6	17.4	1.98	0.11	18	31
98	7.1	20.1	6.4	2.28	7.21	17	36
Canada	**7.1**	**12.8**	**10.9**	**1.00**	**1.00**	**16**	**36**

[1] Month 1 =January 1980, Month 16 = April 1981, Month 36 = December 1982, and so on.

In 1981-82 the most astonishing decline in employment occurred in the mining industry which shed 52,000 jobs, fully 24 percent of the industry total. Being a highly localized industry, the impact of this decline was felt in only a few regions. The structural loss of mining jobs in the Sudbury region was close to 7,000 jobs, in the Province of Alberta 16,600 jobs and in Cape Breton Island, in Rouyn and Noranda, and in the Kamloops-Trail region of British Columbia around 1,000 each. The competitive effect in Sudbury and the latter three regions was a loss comparable in size to the structural effect, whereas in Alberta it was a gain of 12,500 jobs. The reason for this difference is the heterogeneity of this sector: The oil and gas industry was still experiencing fairly buoyant conditions (the tail end of the 1979 OPEC price jump) whereas the rest of the mining industry went into recession. Had the data made it possible to separate the oil sector, the structural effect would have been even larger. Another sector recording a large decline in the downswing phase was the manufacture of durables which shed 150,000 jobs. The structural effects of this decline in the Montreal, Toronto and Hamilton regions were, respectively, losses of 24,700, 37,700 and 15,000 jobs accounting for 34 percent, 57 percent and 50 percent of the total increase in unemployment in that year in those three metropoles.

In 1982-83 the most impressive recovery was that of the forestry sector which increased its employment by 19 percent (+ 12,600 jobs). This structural effect accounted for almost all of the increase in forestry sector jobs in British Columbia (+5,500) and in Ontario and Quebec (+ 7,000).

One of the more interesting sectors in 1983-84 was the small fishing sector. While Canada continued to rack up a modest recovery, fishing lost some 2,200 jobs (7.2 percent of the total). Being concentrated on the east and west coast, this decline was locally serious and contributed substantially to rising unemployment in Newfoundland, and the Yarmouth region of Nova Scotia where a rise in the unemployment rate from 10.1 percent to 13.0 percent was accounted for mainly by this one sector.

6.3 TRADE AND UNEMPLOYMENT

We have shown, in the previous section, that industrial structure played a major role in determining patterns of employment and cyclical unemployment in the 1981-84 period. The final step is to establish a connection between employment levels in an industry and external trade, thereby linking unemployment to international changes. This is the most difficult step, for three reasons. First, of the 12 employment categories, only the first five enter in a major way into Canada's external trade. The other seven categories mainly serve the domestic market. As it happens, it is the first five categories—agriculture, mining, forestry, fishing and hunting, and durables—that experienced the largest employment fluctuations and contributed most to changing levels of unemployment. Second, the data are not directly comparable. Imports and exports are recorded using a commodity classification that of its nature does not match the categories used in the Labour Force Survey. And third, most of the trade data are tabulated for Canada as a whole; only a few aggregate tables identify the province of lading. This sometimes makes it difficult to relate unusual trade performance in a commodity with any particular region, but some tentative connections can be made.

The largest employment decline in 1981-82 was in the mining industry. This was matched by a corresponding decline of 22 percent in the total exported value of metal ores and concentrates (see Table 6.3). Within this group of commodities, major export declines were recorded by iron ore (from $1,465 m. to $1,034 m. or -29.4 percent), copper ores (from $498 m. to $397 m. or -20-3 percent), nickel ores (from $533 m. to $299 m. or -43.9 percent), and precious metals (from $434 m. to $326 m. or -27.2 percent). Indeed, among this group only radioactive ores showed an increase in exports. Production of these various ores is highly localized, and in some cases regional employment decline can be directly matched with trade decline. Labrador's iron ore mines, and the Sudbury region's copper, nickel and precious metal ores are cases in point. Mining in British Columbia also saw a major employment decline from 28,300 to 15,100, but in this case the data make it more difficult to connect specific regional activities and trade.

TABLE 6.2
PERCENTAGE CHANGE IN EMPLOYMENT IN MAJOR CANADIAN INDUSTRIES 1981-82, 1982-83 AND 1983-84 (4TH QUARTERS)

	1981-82	1982-83	1982-84
1. agriculture	98.83	103.81	94.57
2. mining	75.75	101.37	114.28
3. forestry	93.79	118.95	96.08
4. fishing and hunting	100.31	95.63	92.81
5. manufacturing: durables	84.62	105.31	106.37
6. manufacturing: non-durables	91.51	106.84	99.05
7. construction	86.83	98.92	103.81
8. transport and utilities	96.34	96.88	99.12
9. trade	94.98	103.36	104.62
10. finance, insurance, real estate	965.05	107.19	103.02
11 services	101.72	104.48	102.51
12. public administration	102.85	100.11	101.03
all industries	**95.68**	**103.44**	**102.354**

Source: Statistics Canada, special tabulations.

The second sector to experience a major employment decline in 1981-82 was the manufacture of durables. Within the polyglot group of commodities produced by this sector, the main items recording a large export decline were: wood pulp (from $3,221 m. to $2,738 m, or -15.7 percent), primary iron and steel (from $2,315 m. to $1,969 m. or -14.9 percent), non-ferrous metals (from $5,422 m. to $4,807 m. or -11.3 percent), industrial machinery ($2,739 m. to $2,485 m. or -10.2 percent) and agricultural machinery ($885 m. to $651 m. or -26.3 percent). Offsetting these was growth of transport equipment exports from $15,847 m. to $19,460 m., an increase of 22.8 percent. Most of this increase was accounted for by the auto industry, with the exports going almost exclusively to the US under the 1965 Canada-United States Automotive Trade Agreement. Since Canadian imports of transportation equipment declined from $19,713m. to $17,526m. in the same year, there was a remarkable swing from deficit to surplus in the balance of trade in automotive products.

**TABLE 6.3
SELECTED TRADE STATISTICS
1981-84 (MILLIONS OF DOLLARS;
AND FIXED WEIGHTED VOLUME INDEX «1971=100»)**

Domestic Exports		1981	1982	1983	1984
Fish	$ m.	1,481	1,583	1,546	1,572
	f.w.v.i.	152.3	157.0	148.9	150.3
Meat and preparations	$ m.	628	779	701	755
	f.w.v.i.	221.4	259.3	250.2	280.3
Wheat	$ m.	3,710	4,289	4,648	4,710
	f.w.v.i.	124.0	157.9	168.9	168.7
Barley	$ m.	n.a.	886	815	636
	f.w.v.i.	n.a	n.a.	n.a	n.a
Other cereals and preparations	$ m.	n.a	537	528	557
	f.w.v.i.	n.a	n.a	n.a	n.a
Coal	$ m.	1,147	1,269	1,313	1,847
	f.w.v.i.	231.6	233.2	242.8	321.3
Petroleum	$ m.	2,505	2,729	3,457	4,390
	f.w.v.i.	21.6	28.4	39.0	48.4
Gas	$ m.	4,370	4,755	3,958	3,886
	f.w.v.i.	82.5	84.7	77.8	81.8
Iron Ore	$ m.	1,465	1,034	972	1,112
	f.w.v.i.	116.8	78.7	77.4	91.3
Other metal ores, etc.	$ m.	2,619	2,158	1,924	2,558
	f.w.v.i.	n.a	n.a	n.a	n.a
Lumber	$ m.	2,989	2,912	3,964	4,254
	f.w.v.i.	149.9	143.3	170.4	185.7
Wood pulp	$ m.	3,820	3,221	3,058	3,908
	f.w.v.i.	138.3	118.0	131.8	136.3
Newsprint paper	$ m.	4,325	4,086	4,005	4,784
	f.w.v.i.	118.1	100.2	105.4	113.8
Autos	$ m.	n.a	7,358	9,573	13,890
	f.w.v.i.	n.a	n.a	n.a	n.a
Other transport equipment	$ m.	n.a	12,113	14,208	18,455
	f.w.v.i.	n.a	n.a	n.a	n.a

n.a — not available
Source: *Summary of External Trade* (65-001) Tables 8 and X3
Exports: Merchandise Trade (65-202) Table 3
Special tabulations by Statistics Canada.

Thus there is some evidence of trade decline matching employment changes, but the picture in the durables industry is murky, not least because of the widespread and sometimes drastic restructuring that was going on in the sector during the early stage of the recession. The resulting widespread work force reductions make it difficult to connect employment with trade because output levels did not necessarily change in proportion to employment changes.

In the following year, 1982-83, the largest employment increase was achieved by the forestry industry. On the trade side, exports of lumber rose dramatically from $2,912 m. to $3,964 m. or +36.1 percent, whereas a small decline was recorded in the export of pulp and newsprint. Since British Columbia specializes in lumber, Ontario and the Atlantic Provinces in pulp, while Quebec produces both pulp and lumber in volume, one would anticipate these export trends to be reflected in regional employment changes. In this case, they were: employment in forestry in Ontario and the Atlantic Provinces remained almost static, whereas the two main lumber producing provinces—British Columbia and Quebec—showed healthy gains. Two other sectors, non-durables, and finance, insurance and real estate, also recorded rapid employment growth from 1982 to 1983. However these activities mostly serve the domestic market so that it is not meaningful to relate trade and employment patterns.

In 1983-84, the two sectors with the strongest growth record were mining and durables. Exports of metal ores and concentrates grew by 26.6 percent in dollar value, and the fixed weighted volume indexes for coal, iron, nickel and copper ores, asbestos, crude petroleum and natural gas also grew substantially. Among durables, the largest work force is in transport equipment, and here too exports (predominantly vehicles and parts to the US) grew substantially in value and volume equivalent. In both sectors there is a correspondence between an increase in exports and regional employment growth. Increased exports of metal ores and concentrates, and oil and gas can be matched with work force increases in the mining sector of 3,300 in the Calgary region and 5,500 in northern Alberta (both mainly oil and gas), 3,300 in the Rouyn-Noranda region (precious metals, copper, zinc), and 1,600 in the Sudbury region (nickel, zinc, copper and radioactive ores). In the durables sector, major employment increases were recorded in Central Canada, notably the Montreal region (+9,700 jobs), the Toronto-Oshawa region (+8,000) and Kitchener-Waterloo (+12,300). All these regions have major plants in the transportation equipment sector, and though a direct causative connection cannot be made with the present limited data, it seems highly likely that the growth of exports in this sector generated new jobs in regions specializing in in manufacturing durables.

6.4 CONCLUSION

Very little has been written on the subject of trade and unemployment, yet it seems highly probable that in open trading economies trade performance will have a direct effect on economic activity, employment and unemployment. This is, in effect, the heart of the economic base concept, and its Canadian variation known as staple theory (Bertram, 1967; Watkins, 1977). Compare, for instance, Australia and Canada on the one hand with the United States on the other. All three countries have large resource endowments, but in the cases of Canada and Australia the populations are small so that the resources are available for export. In 1983, Australian exports accounted for 14.8 percent of GDP (despite protectionism), while in Canada exports constituted 25.8 percent of GDP. The United States, in contrast, exported only 7.7 percent of GDP in 1983, reflecting a much higher degree of economic closure. These differences suggest that in Australia, and in Canada in particular, regional unemployment is more likely to be affected by external trade than in the United States.

This broad and self-evident theme is not easily demonstrated in practice, however, because it is difficult to reconcile merchandise trade data (recorded by commodity) with employment data (recorded by industry). The analysis has, in the first instance, related unemployment to regional industrial structure using a modified shift-share model. It was then found that the performance of the specific industries in which a region specialized had a significant influence on levels of unemployment in that region. The final step linking industry to trade was much more tenuous, and the evidence more circumstantial. Nevertheless, a consistent picture began to emerge, connecting regional export trade performance to regional unemployment.

Though this paper is not concerned with policy issues, it is important that one direct policy implication of the discussion be unambiguously identified. It is common, when policies to locate industries and jobs in problem regions are propounded, that the effects on regional exports are considered. But the reverse case, it would seem, is rarely considered. Yet trade agreements have an impact on regional employment and unemployment. Specifically, if the preceding analysis is correct, then the Free Trade Agreement between Canada and the United States will affect regional unemployment patterns in Canada. However, though much has been written on the expected effect of the agreement on productivity and efficiency in Canadian industry, the consequences for regional labour markets seem to have been scarcely considered. Yet the connections between international trade and labour markets demonstrated here, and also by Raynauld (1987), suggest that it should be a priority to monitor the consequences of the agreement for regional patterns of employment and unemployment in Canada.

7. LABOUR MARKETS, HOUSING MARKETS AND THE CHANGING FAMILY IN CANADA'S SERVICE ECONOMY

David Ley and Caroline Mills

Perhaps it should not be a surprise that in a century labelled by philosophers as the age of analysis, and in a society in which specialization has been taken to extreme lengths, that social science has opted more often for fragmentation than for coherence, for dividing rather than for integrating. So we find an economic geography of production, of the workplace, and a social geography of consumption, of the homeplace, with only the most slender of literatures to bring them together, other than the literature of the sledgehammer where one is reduced to the other. We find an increasingly sophisticated literature on labour markets existing largely in isolation from an increasingly sophisticated literature on housing markets. As for family status, it has scarcely entered the domain of human geography at all, other than in some feminist research and in the limited manner of ecological models of urban structure. Yet arguably, work relations, home relations, and family relations *together* not only define a more theoretically valid view of society and space in the city but also define a more existentially valid view of the meaning of urban life to its workers-cum-residents.

The biography of any household shows the interplay of production and consumption, of workplace, homeplace, and family relations at the levels of both context and meaning, the objective and the subjective. However, such a synthesis of people and place, though desirable, is by no means an easy task to accomplish and the present chapter falls well short of such an achievement. But in drawing together several linked projects to outline the interrelated realms of production and consumption in a contemporary Pacific Rim city, it will begin to point to the possibilities of such a research agenda.

Curiously, there has often been quicker recognition of these interdependencies in the developing world around the Pacific Rim than in advanced societies like Canada and Australia. In an ambitious project, McGee (1985) is tracing the onset of a multinational, corporate economy in Malaysia not only in patterns of economic development, but also in changes within the household as family members (usually young women) enter the industrial work force and their attitudes toward the family and personal consumption begin to be redefined, sometimes with competing loyalties resulting. In a similar spirit, Christopherson (1983) has examined the adjustments of Mexican households to the opportunities and tensions introduced by employment in the American corporate economy which is encroaching across the Mexican border in search of cheaper labour and less rigorous government regulation.

A third study, close to our own point of departure, is Geraldine Pratt's examination of a Vancouver suburb, where she traced the interdependencies between attitudes toward housing tenure, employment status and family structure (Pratt, 1984, 1986). She found a disproportionate tendency for suburban families who were tenants to include a household head who was a self-employed, small businessman, and who had made a decision to defer homeownership in order to concentrate scarce financial resources into his economic ventures; some indeed had used capital gains from prior homeownership to boost resources for the business. Other tenant households were foregoing homeownership on different grounds. Having given priority to the presence of a spouse at home during the years of child rearing, they were unable to enter the home-buying market on a single household wage. A third tantalizing finding suggested a relationship between housing tenure and union activism, with the responsibilities of a mortgage and home ownership tempering the level of union militancy. In these ways, and others, Pratt demonstrates that bonds between the realms of production and consumption, and between work, home and family life which are contingent and reciprocal.

Our own concern is not with the suburbs, but with the inner city as a housing market intimately bound to the changing character of the downtown area as a labour market. Moreover, in the restructuring to what has sometimes been called a post-industrial society, shifts in housing and labour markets are closely bound up with a redefinition of family relations. Our discussion will now seek to elucidate some of these linkages in the Canadian inner city, and more specifically within the city of Vancouver.

7.1 THE RISE OF SERVICE-BASED EMPLOYMENT

One of the central premises for the emergence of a post-industrial society is the fundamental transition of jobs from primary and secondary to tertiary and quarternary categories, a transition proceeding often faster than the contribution of services to the gross domestic product (Bell, 1976; Ley, 1983). In Britain, for example, the loss of over a million industrial jobs from 1966 to 1974 was countered somewhat by the gain of 700,000 new jobs in services (Sant, 1978). These trends are continuing, as an exhaustive current survey anticipates the loss of a further 665,000 primary and secondary jobs in Britain between 1985 and 1990 and the addition of 540,000 new positions in services (IMS, 1986). Similar shifts are occurring in Canada, where even in a staple-led economy like British Columbia's, some 70 percent of employment and gross domestic product (GDP) was accounted for by the service sector by 1981-82 (Waters, 1985).

If the service sector is disaggregated, the trends are even more striking, for job growth is occurring disproportionately in the senior white-collar categories, including managerial, administrative, professional and technical positions, which have shown far greater resilience than other sectors during the recession years of the early 1980s in British Columbia. This sector added some 40,000 jobs during the 1981-84 period while all other occupational categories shed some 90,000 workers; over the longer 1975-84 period it was responsible for two thirds of job growth (Daniels, 1985). As a result, by 1984, the quaternary work force in British Columbia as in other regions accounted for more than a quarter of the employed labour force.

A second important disaggregation of the service economy is a spatial one, for there is a significant geography to employment change. Within provinces most growth is within cities, and within individual cities most growth is also spatially concentrated. In Vancouver, for example, one third of recent new metropolitan jobs have been concentrated in the urban core (City of Vancouver, 1984), while 90 percent of new city jobs have fallen in the industrial categories of finance, insurance, and real estate (the FIRE group), community, business, and personal services, and the public sector. Survey results indicate that professional and technical positions will continue to expand by between 5 and 8 percent in greater Vancouver to 1990 at least, while low-skilled jobs and clerical staff will be in reduced demand (VEAC, 1984). Of particular interest is the growth of the producer service categories (Ley and Hutton, 1987). The city issued 1,800 licences to firms with downtown addresses engaged in business and professional services in 1971, and 3,500 in 1981. Firm growth in some fields is remarkable; a net increment of over 1,200 legal firms has taken place between 1960 and 1983; management consultants have increased eleven fold, and whereas there were no data processing firms listed in 1960, by 1983 there were 124. All of these firms are concentrated in the urban core; over 80 percent of metropolitan legal firms, the most spatially biased of the group, have a downtown location.

There is one other trend concealed by the aggregate figure that requires comment. Two of every three new jobs created in British Columbia during the 1970s is held by a woman; half of the staff in downtown head offices is female. Traditionally, there has been a division between male management and professional staff in head offices, and female clerical workers, leading to a dual labour market stratified by gender, wages, and career prospects (Stanback and Noyelle, 1982). Clerical staff have come under a variety of pressures in recent years. First, inflationary costs of shelter in the inner city are displacing them from this historic residential zone of less well-paid service workers; they are moving to the suburbs and therefore have further to go to work (Ley, 1985a). Second, office automation is putting clerical jobs at risk. A survey of the largest head offices indicated they had shed close to 13 percent of their clerical staff between 1980 and 1985 (Hutton and Ley, 1987). Other data indicate this trend will continue with a 5 percent loss from 1984-89 (VEAC, 1984). As a result clerical unemployment in Greater Vancouver exceeded 10 percent by 1984, which was twice the rate of the quaternary sector.

Other significant changes have begun to take place. Upgrading and the addition of new skills have occurred to a greater degree in clerical work than in other segments of the services labour force (VEAC, 1984). Perhaps more important, women are entering quaternary-level employment at an accelerating rate. Despite the significant decrease in clerical staff (which is made up almost entirely of women) in Vancouver head offices, a small increment in the female share of the work force has taken place in the past five years. Women are clearly beginning to move into more senior employment positions; three quarters of major corporations surveyed in Vancouver reported that this development had indeed taken place in their own head offices.

Thus economic development in BC (in particular new job formation in the highest status occupations) has been concentrated in a small area, metropolitan Vancouver, especially its central district. This focussed growth has clear implications for the restructuring of the built environment. First, the concentrated pattern of workplace growth has led to a downtown building boom, and a tripling of office space in the urban core between 1967 and 1984. Second, the fact that a significant proportion of these employees are relatively well-paid professionals and managers introduces a distinctive presence to the urban housing market. Third, the fact that women comprise a growing share of this cohort introduces new possibilities for household formation, family relations and thus the type and location of housing units that will be sought.

7.2 ECONOMIC DEVELOPMENT AND THE BUILT ENVIRONMENT

We can now begin to draw together our argument concerning the linkages between labour markets and housing markets. Fuelled by a staple economy of forest products and minerals exported to world markets, the British Columbia gross domestic product (GDP) has grown consistently in the post war period, by some 6 percent per annum in the 1960s and 5 percent in the 1970s. As recently as the 1971-81 period total investment in British Columbia tripled, though it then fell in each of the next three years. Indeed in the depth of the recession in 1982 corporate pre-tax profits plunged to only 27 percent of the peak recorded in 1979. In the 1980-84 period provincial GDP increased by only a little more than one percent a year.

The performance of the staple-based provincial economy is closely correlated with the growth of jobs in the service sector. The category of community, personal and business services, by some distance the largest service grouping, has been expanding in locked step with moves in the provincial GDP. The GDP growth peaks of 1965, 1969, 1973, 1977 and 1980 all reappear on the graph of service growth, as do the troughs of 1970, 1975, and 1982-83. Both of these trends are themselves highly correlated with the addition of new office space in downtown Vancouver. A model has been developed predicting the expansion of office space in the central business district (CBD) as a lagged function of the growth of the provincial economy (City of Vancouver, 1985). The phases of the business cycle in the provincial economy are said to be transmitted to downtown office construction with a four year lag, the time it takes a developer to make a decision, assemble a site, acquire financial and planning approval, and complete construction. According to this proposition, the peaks in the growth of the provincial GDP in 1965, 1969, 1973, 1977 and 1980 are communicated to the pulses of downtown office construction in 1969, 1973, 1977, 1981 and 1984.

A fourth indicator, Vancouver house prices, follows a similar cycle. The average house price inflated rapidly during 1972-75, coinciding with the peak in the business cycle of the early 1970s. A softer market followed in the middle years of the decade, but by 1979-81 with renewed provincial GDP growth there was precipitous inflation, with some residential properties doubling in value in 12 months. In 1981 service employment underwent spectacular growth of over 10 percent, and in the spring of that year residential values were at their peak, and the city had the most expensive house prices in North America. Then with the recession, prices slumped by up to one third and a generally soft residential market matched the depressed state of other indicators through 1982-84.

So the statistical associations between economic growth, new job creation, and consequences for the housing market are fairly clear in aggregate form. But our objective is more specific. What is the narrower set of relations binding the rise of quaternary employment downtown, the inner city housing sub-market, and new household and family form, not only in terms of a network of statistical associations, but also at the level of cultural values invoked by individual and household decision-makers? We shall attempt to make some progress in exploring these relations as we consider the process of middle class settlement of the inner city popularly known as gentrification.

7.3 CHANGING PATTERNS OF WORK, HOME AND FAMILY LIFE: INNER CITY GENTRIFICATION

Gentrification is a loaded term with at least two nuanced meanings. First, it conjures up the process of class succession in the inner city, with the settlement (often, more properly, the resettlement) by the middle class of residential areas which have filtered down to be affordable to poorer households. Consequently neighbourhood change is accompanied by the displacement of these households. Commonly, the in-migrants are called the new middle class, or more simply the new class, identifying them with a theoretical literature which discusses the emergence of an objectively new grouping in twentieth century market societies (Gouldner, 1979; Bruce-Briggs, 1979), a cohort which coincides to a considerable degree with the quaternary employment category used earlier. In short, gentrification and the advanced service economy of the downtown are two sides of the same coin, the intersection of a labour market and a housing market. Within six of Canada's largest cities, quaternary workers living in the inner city rose in number by one third from 1971 to 1981, while non-quaternary workers declined by 20 percent (Ley, 1988).

But gentrification also evokes status, a set of values and activities which define a distinctive lifestyle. In some parts of the world, gentrifiers are known (derogatorily) as "the trendies," and this label identifies an attention to style, fashion and tasteful living which has been perceptively captured by several creative writers including Hugh Garner in Toronto (1976), Jonathan Raban (1974) in London, and Sherman Snukal (1982) in Vancouver. Two particularly prominent elements of the lifestyle will be outlined and then developed in more detail in this chapter.

First, as we have suggested, there is a distinctive approach to consumption, the pursuit of objects for their symbolic significance rather than for their functional value alone. Chasing as well as leading this market, the successful designer has to include "new inspirations deriving from irony, curiosity, surprise and friendliness" (Branzi 1984, p. 148). Typical is the retail landscape of Fourth Avenue in Vancouver's gentrified Kitsilano neighbourhood. The street has been "making a comeback as a recreation and shopping area for the upwardly mobile condominium set. Trendy restaurants, clothes shops and other speciality stores are springing up like mushrooms after a spring rain" (Gee, 1978). On the other side of the continent on the Atlantic Ocean, we find in Spring Garden Road, on the downtown side of the gentrifying South End of Halifax, an identical symbolic repertoire. New shops are heralded for a market "looking for an alternative of unique buys over mass merchandise...from unique books, to trendy canvas clothing, right on up into high fashion for men and women. We know that our boutique owners share the 'small is beautiful' philosophy and we are ready to offer Metro shoppers personalised service" (advertisement in Ward One Community News (Halifax), March 26, 1984). This designer culture is expressed in residential landscapes, both in the fastidious restoration of older properties with valued architectural signatures in cities like Toronto and Melbourne (Holdsworth, 1983: Jagar, 1986), and also in the multivalent designer cues of post-modern architectural styles so evident in inner Vancouver (Mills, 1988).

A second characteristic of the lifestyle is attitudes toward family status. A decision either to delay or reject the conventional model of the nuclear family is a common feature of inner city gentrifiers, and allows additional discretionary income for the household. Responding in part to changing labour market opportunities, over half of all Canadian women were employed in 1981, a 4O percent increase over the 1971 level. As we have seen, in centre city they include a growing cohort of women in senior positions in the service economy. Either as single male or female salaried workers, or as dual salary households, decisions concerning a preferred family type interact directly with attitudes towards both careers and also location within the housing market. This situation is particularly true for gay households, a small but conspicuous group among the gentrifiers (e.g. Castells, 1983b). Their antipathy toward the conventional nuclear family repels them from the suburbs with their conformist family model, toward the more plural options of the inner city. We are seeing here the engagement of production and consumption relations to produce a changing human geography in the inner city. Some years ago a survey in Chicago identified careerism and consumerism as guiding ideologies of middle-class households living in the central city, and familism (in the traditional sense) and communalism as life choices for suburban households (Bell, 1958). The process of gentrification has arguably polarized these competing ideologies still further in the intervening period.

It remains to present some empirical evidence to support the arguments outlined above. We shall proceed in two stages: first, employing a more formal methodology, and emphasizing the class dimension of gentrification, we shall look for statistical associations between labour markets, housing markets, and family status in the Canadian inner city; second, with a qualitative methodology and emphasizing the status dimension the same question will be addressed at the micro-scale of a specific Vancouver neighbourhood.

Correlates of Gentrification in Metropolitan Canada

Gentrification was defined as an appreciable upward shift in inner city social class recorded between the 1971 and 1981 censuses. An index of social class change (or gentrification) was computed for the inner city districts of the 22 metropolitan areas recognized by the Canadian census in both 1971 and 1981 (for additional details, see Ley 1985b, 1986). Correlations were computed between the gentrification index and a set of independent variables (Table 7.1).

TABLE 7.1

SELECTED CORRELATIONS AGAINST GENTRIFICATION

22 CANADIAN CENSUS METROPOLITAN AREAS, 1971-81

Metropolitan office space per capita of CMA Population, 1981	0.65
Metropolitan job growth, 1971-81	0.46
Percent non-family households in CMA, 1981	0.39
Change in percent non-family households, 1971-81	0.45
Female participation rate in CMA labour force, 1981	0.39
Index of resident satisfaction with CMA, 1978	0.54
Perceived environmental quality of CMA, 1978	0.54
Art galleries per 10,000 population in CMA, 1981	0.46

Source: Ley (1986). Were significance tests applied, correlations of r = 0.54, 5 = 0.42, and r = 0.35, would provide thresholds for significance levels of .01, 0.5, and .1 respectively.

One of the earliest studies of the restructuring of the American inner city showed that during the 1960s in the 20 largest cities in the United States a strong association existed between then incipient gentrification and the presence of an economic structure oriented toward office-based services (Lipton 1977). This argument has been restated, though incompletely tested, more recently (Gale, 1984; Berry, 1985). It clearly holds in Canada also, for among the 22 cities the leading six on the gentrification index from 1971 to 1981 were Halifax, Ottawa, Vancouver, Victoria, Calgary and Toronto, regional or national centres of advanced services in the private sector and/or government. The six cities at the foot of the list were Thunder Bay, Winnipeg, St. Catharine's-Niagara, Sudbury, Saint John, and in last place, Oshawa, a group dominated by a primary and secondary economic base. Indeed the variable with the highest correlation (of 35 tested) against the gentrification index was the amount of metropolitan office space per head of population; a more moderate correlation existed against the rate of new job formation in each city (Table 7.1). These results once again underline the close associations between labour and housing markets. Further analysis at the census tract level in six of the largest cities led to the additional finding of a significant association between middle class resettlement in an inner city tract, and the tract's proximity to a university or hospital, important employers of the new middle class (Ley, 1988). Turning to family status, modest positive correlations exist between gentrification and measures of family status, including the census category of non-family households (Table 7.1). So too a measure of female participation in the metropolitan labour force shows a positive correlation with the incidence of gentrification, completing the circuit of workplace, homeplace, and family relations.

One other dimension identified earlier shows up in the analysis. There is considerable evidence that the quality of the natural and built environments is an important factor in the gentrification process. An American study, for example, showed that almost 90 percent of neighbourhoods in a range of cities undergoing gentrification contained some significant environmental amenity or heritage architecture (Clay, 1979). The Canadian data similarly show some robust associations between measures of environmental quality and middle class resettlement, both at the level of the urban system (Table 7.1), and at the more sensitive level of census tract changes in the major cities. At this intra-urban scale, a highly significant correlation existed between the upward social status change of a census tract and proximity to a regional park or accessible waterfront.

There are clearly some limitations to this form of analysis with its aggregate data and rigidly specified variable definitions which no doubt contribute to some slippage in the strength of associations. Nonetheless the interlocking structure of production and consumption relations shows up clearly in the results. Perhaps a more persuasive depiction of how this structure unfolds in everyday life may be offered by a qualitative methodology, with a more nuanced account of how specific inner city households integrate workplace, homeplace, and family relations.

Genre de Vie in a Gentrified Neighbourhood

The extremes of inner city social change in Vancouver are epitomized by

Fairview Slopes. Comprising about 20 city blocks, a five minute drive from downtown, this neighbourhood, along with the publicly developed False Creek residential district, comprise the census tract with the city's highest index of social class change (gentrification). Under piecemeal development by private developers, the character of Fairview Slopes has dramatically changed in the last 15 years.

An early planning document described the neighbourhood as "bordering on a rural lifestyle caught up in the urban animal of Vancouver" (Elligott and Zacharias, 1973). At that time the population comprised a mix of working class people with communal households pursuing "alternative lifestyles." Dilapidated single family houses, converted to rooming houses, were interspersed with some light industrial and warehouse uses. A zoning change in 1976, permitting floor space ratios of up to 1.5, boosted property values and precipitated a flurry of activity associated with multiple house sales and redevelopment (Lum 1984). This peaked in 1980, slowed with the recession, and increased again in 1985 and 1986.

Today, Fairview Slopes is a stylish district of medium density townhouses and condominiums in the post-modern design idiom, with a few renovated older houses and some remaining commercial buildings. One architectural critic described it as "one of the few neighbourhoods in Vancouver with an urban feel" (Gruft, 1983). This quality is promoted by the marketing for new condominium developments which are said to offer "city living at its best," and "all the conveniences of an urban lifestyle." In line with the findings of the quantitative analysis of gentrification, Fairview Slopes is rich in amenity value (with spectacular views and close to leisure facilities) and is also well-placed with respect to "new class" employment downtown and in the adjacent Broadway medical, office, and retail corridor.

Our task is to relate how the elements of gentrification — labour market, housing market, cultural values, and attitudes towards the household — are experienced in everyday life. In addition to interviews with residents, architects and developers ,we have drawn on the lifestyle imagery expressed in advertisements for new buildings (Mills, 1989). Both landscape "consumers" (buyers or renters) and "producers" (planners, architects, developers) play an active role in constructing an ideology of the "good life" which requires a particular kind of residential setting. A central theme of that ideology concerns occupational status.

Census data locate residents of Fairview Slopes within the changing urban labour market. In 1971, of the Fairview Slopes labour force stating an occupation, 14 percent were in the new middle class, claiming managerial, administrative, professional and technical occupations. Clerical, sales and service occupations accounted for 54 percent of those with stated occupations. By 1981, 44 percent of the labour force giving an occupation had new class type occupations. Employment in clerical, sales and service categories remained high (38 percent). These remained especially important for women (54 percent of the female work force) although a significant proportion of women (38 percent) held new class occupations. These patterns reflect a highly educated population and in 1981, 43 percent of adult residents had received some university-level education.

The upward shift in social status associated with the growth of senior white-collar occupations in the central city contributes to a social class change in the inner city housing market. In a 1980 survey (Fujii, 1981) Fairview Slopes residents stressed proximity to workplace as a reason for moving here, listed second, after beautiful views. A central location especially facilitates the two journeys-to-work of dual-career households. In addition, many new class jobs demand flexibility in work schedule and a willingness to sacrifice "free time" for work, reinforcing the desire to live where, to quote one resident, "I'm close to work. ..On the weekend, quite often I'll drive down there and work!" Compact condominiums with no lawns to mow are preferred because they demand little dedication to maintenance activities.

Preferences for a neighbourhood type also respond to the status element of career. The occupations of the Fairview Slopes population typically carry with them an element of prestige. The neighbourhood's reputation as a home for successful young professionals is a theme of the marketing literature. Living in "prestigious Fairview Slopes" is said to make a "statement about societal achievement," Fairview offers "a style of living you've worked hard to attain...all the prestigious amenities you've sought. And deserve." Not only does it reflect career success but (we are led to believe) it may even enhance career prospects; this is "a friendly village of people on the way up," and "perfectly adapted to an upwardly mobile lifestyle." "Can you afford not to be living here?" asks one advertisement.

Careerism and consumerism are the typical lifestyles of Fairview Slopes. They converge through the status element of career; indeed, Bell (1980) has suggested that the term "new class" is better applied to a status group with a style of conspicuous consumption, rather than to an occupational class. A number of Fairview Slopes residents experienced a fuzzy boundary between what are conventionally perceived to be separate spheres of life; to quote: "There aren't any distinctions between my work and the rest of my life, they all meld into one homogeneous kind of existence." One architect describes the synthesis of lifestyle imperatives which transformed the neighbourhood: "You had to have the right kind of job and drive the right kind of car and live in the right kind of inner city unit...it's become the yuppie lifestyle." Of course, their affluence permits them to indulge in the rich variety of consumption amenities offered in the city, and to outbid other groups for desirable central locations. However, consumption style itself is not reducible to economic class. Our case study warns against fragmentary analyses which focus on production or consumption alone.

Household type is another significant element of the Fairview Slopes *genre de vie* . This is one of Vancouver's most demographically biassed neighbourhoods outside downtown and its immediate environs. In 1981, over 40 percent of adults were single, and 10 percent were divorced (not remarried); 55 percent and 33 percent of households comprised one and two persons respectively. There are few children. Many couples are of the dual-career type, both partners pursuing careers which require a high level of commitment. The proportion of residents who can be classified as "homemakers" fell from 13 to 3 percent of those on the 1975 and 1985 voters' lists.

Household types in Fairview Slopes reflect changing attitudes towards the family and gender roles, but they are also shaped in response to consumption style and career activities. Postponing marriage or childbearing frees resources for other activities, an advantage for one resident in that "you have nobody to answer to, you can stay at the office as long as you like." Producers of Fairview Slopes housing target this market group. According to developer Andre Molnar, "The profile is single, male or female, business-oriented, well-educated, reasonably affluent people, with a large interest in recreation and entertainment, or people who are newly married or living together....I cannot think that we sold to anybody with a family." Smaller units built since the recession have proved particularly attractive to the single woman, and pictorial advertisements may feature such a consumer. The theme of non-traditional gender roles also appears in a cartoon advertisement showing a woman announcing to her husband that she has bought *herself* a new townhouse. A willingness to compromise conventional family life has both costs and rewards with respect to levels of satisfaction, as explained by this dual-career couple: "We have...an egalitarian kind of life where both people are striving towards certain goals...we have a fair amount of independence from one another, our work life makes each other's lives harder, but there's a parallel theme there which mutually aids you in a sense that the other person's doing the same thing." This moral support is typically upheld by the sharing of household tasks.

The Fairview Slopes *genre de vie* has a distinctive landscape expression. Housing producers respond to changing lifestyles and preferences, but their product also acts back by constraining or facilitating residents' lifestyle projects. The size and character of the housing unit is appropriate to the everyday activities of the residents. Condominiums built on a number of different levels are appropriate for young and active persons; the separation of living space gives partners physical and psychic room to pursue their separate projects. Features such as bedrooms on balconies overlooking living rooms make some units suitable only for childless households. The distinctive architectural style also seems attuned with residents' lifestyles. Social status is boosted by living in this neighbourhood which has one of the highest densities of award-winning residential architecture in the nation. Post-modern designs are rich in significance for the designer culture, including the use of elements such as classical columns and triumphal arches, which signify the status of the new class as a new elite. To incorporate cues from other contexts, as varied as Georgian London or Mediterranean Spain, also implies the cosmopolitan sophistication of its worldly occupants. The attraction for people with unconventional gender roles is legible in some facades with bedroom colours and "San Franciscan styling"; "the rounded shapes...the sculptured buildings, are very attractive to the women and to the gays," claims one developer who has tapped these submarkets.

In Fairview Slopes a distinctive landscape reveals the lifestyle ideology of its residents. The interplay of the realms of culture, the home and the economy is the very nature of gentrification, and can only be understood through a synthetic approach.

7.4 CONCLUSION

In this paper we have argued that restructuring is an economic process, but that it is also more than this. Restructuring is implicated in the realms of both production and consumption, in the creation of new landscapes and new *genries de vie* in short in the shaping of a new human geography which involves changing and interdependent relations of work, home, and family life. We have attempted to integrate these domains in our study of the changing Canadian inner city. Clearly this paper offers only a preliminary synthesis, but it is suggestive of a more comprehensive and successful integration of a range of conceptual heuristics employed in the exploration of late twentieth century society in Canada, Australia, and other advanced market societies. Heuristics such as the post-industrial society, the service economy, the new class, changing gender relations, the culture (and politics) of consumption, and post-modern design styles, are by and large found in separate literatures. Our argument is that they belong together, and that it is the task of a synthetic human geography to bring them together.

8. THE IMPLICATIONS OF INTERNATIONAL CHANGE FOR REGIONAL WORK FORCE STRUCTURES AND CHARACTERISTICS IN AUSTRALIA

John McKay

An important problem facing geography is to explain the complex ways in which changes at the global level have an impact on characteristics at the local or regional level. We need to understand both the importance of world systems influences and the ways in which these forces intersect with local interests. As Timberlake argues in his introduction to a volume on urbanization in the world economy:

> The claim is not that world-system processes determine everything. Rather, the fundamental lesson is that social scientists can no longer study macrolevel social change without taking into account world-system processes. Specifically, processes such as urbanization can be more fully understood by beginning to examine the many ways in which they articulate with the broader currents of the world-economy that penetrate spatial barriers, transcend limited time boundaries, and influence social relations at many different levels (Timberlake, 1985, p. 3).

Thus, the division of labour at the national or international level does have a significant impact upon the urban system. Moreover, recent studies suggest that with the decline in the importance of non-labour factors of location, the geography of labour is assuming a unique and vital role in shaping the processes of economic change. It is the purpose of this chapter to highlight some major features of the changing geography of labour in Australia.

First, I will consider the open nature of Australian society and the ways in which external influences have been important and continue to influence the Australian economy and society. Second, the legacy of Australia's settlement history will be considered briefly in relation to some characteristic features of the regional systems, in particular the dominance of a few large cities, relationships between urban and rural areas, and the balance between the states. Here, the emphasis will be on the genesis of institutions which continue to shape the structure of regional labour markets. Third, recent changes in the geography of employment will be analysed, attempting to isolate where possible the impact of changes taking place in the Pacific Basin. Finally the nature of external pressures likely to face the country in the next two decades will be considered, and some possible regional outcomes outlined.

8.1 AUSTRALIA AS AN OPEN, DEPENDENT SOCIETY

The problem of separating the local from the global, the autonomous from the imposed, is made doubly difficult in Australia since almost all local institutions, and even modes of thought, have been derived from overseas models. What appears as local initiative may sometimes be a delayed copying of overseas fashions or trends. Thus insights into Australian life are gained not only by asking whether local initiatives are important but by trying to identify the impact of external influences coming from different sources. In his provocative book, *The Australian Dilemma*, Bruce Grant (1983) has suggested that modern Australia can be understood through the interplay of three influences: those from Britain, the United States and Asia. Much of the change that has taken place since the Second World War has been the result of the replacement of predominantly British institutions by those drawing their inspiration from the United States. The tensions of the 1980s result from a challenge to both of these traditions by influences from Australia's Asian neighbours. Yet it is important to understand that while Australians have revered American and British models, they are in many cases resisting the winds of change blowing from the Pacific Basin. The values of Europe and North America have shaped Australia's history and institutional base, but the Asian challenge to these old structures is unwelcomed:

> Australians have complained about 'isolation' and the 'tyranny of distance', but the complaint was made in respect of Europe and the United States, not of their near neighbours in South-East Asia and the Pacific, or even in

> the wider reaches of Asia, such as Japan, China and India. On the contrary, for some Australians, whatever distance could be placed between themselves and Asia appears to be an asset, a kind of negative defence. Here is the double dilemma of Australian existence. The dilemma of Australian nationhood is the desire to be a nation, while lacking the will and the capacity to defend the national territory. The dilemma of Australian civilization is that Australia is white, capitalist and Christian in a part of the world subject to powerful Asian influences. Cherishing western values, Australians have become intellectually and materially dependent on the power centres of the western world to protect them from Asia, thus inhibiting the growth of the Australian nation. Until Australians seek for themselves a new form of western civilization, their nationhood is crippled (Grant, 1983: pp. 4-5).

Lest it be unclear just what all of these ideas have to do with the regional geography of labour, I want to argue in the rest of this paper that most of the features of this geography have been inherited from a willingness to adopt outside institutions and a dependent political and economic role in the world. Contemporary changes are a belated recognition of pressures coming from the new world system, and often manifested in the Australian context as influences emanating from the Pacific Basin.

8.2 THE GENESIS OF THE AUSTRALIAN URBAN AND REGIONAL SYSTEM

Timberlake (1985) has identified four secular trends that have characterized the world system since its inception, and which have had an enormous impact at the regional level. First, the integration of new populations and countries into the world economy; second the intensification of commodity relations; third the increasing power of states over their populations; and, fourth, the increasing size of economic enterprises in terms of the amount of capital and numbers of workers controlled by a single firm.

These forces, operating as they did in the early settlement of Australia, significantly rather late in Europe's transition to industrialism, have shaped the geography of the urban system and of labour in several significant ways.

The most obvious way relates to the extremely high degree of urbanization. Australia has always been one of the most urbanized nations in the world, a point stressed some time ago by Butlin (1964) who argued that "the process of urbanization is the central feature of Australia's economic history, overshadowing rural economic development and creating a fundamental contrast with the economic development of other 'new' countries."

A number of writers have made a link between the role of Australia in the international division of labour and the emergence of a high degree of metropolitan primacy. Three basic causes of primacy have been put forward in the literature. The first has stressed the needs of the colonial system. Empires have traditionally been controlled by large cities which have become the focuses for colonial interchange. The second theory has emphasized the export dependency of the colonial situation, emphasizing distortions in the local economy that result from the orientation of the colonial economy to markets in the metropolitan country. The third view has stressed that metropolitan primacy is a sympton of rural collapse, the decline of the old functions of the rural areas highlighting the role of the cities (Smith, 1985). In the case of Australia, Clark (1985) has suggested that it is important to examine not only the functions that were represented in the economy at various points in history, but also the precise forms which these functions took. He argues that the *nature* of Australia's interaction with the outside world was more formative for the development of the urban system than was the *intensity* of these interactions.

This point can be taken a stage further by examining the occupational structure of the emerging metropolitan areas. The cities grew essentially as extensions of the European system and their hinterlands were always oriented towards commercial activities. The cities themselves developed as commercial or mercantile cities, what McCarty (1970) has called "pure products of capitalism." Services have always been the most important providers of jobs in Australian cities. By 1901, more than 50 percent of the national labour force was employed in tertiary activities. The important role of services was partly explained by the fact that manufacturing industry was slow to develop in Australia, to some extent because of the small and dispersed nature of the market and the existence of tariff barriers between the colonies which prevented the emergence of a viable national market. Perhaps more important was the dependence on British exports, an integral part of the colonial system. It was not until the First World War when the cutting of supply routes to Britain forced a degree of self-reliance that Australian industry became firmly established. This process was taken a stage further during and immediately after the Second World War.

The role of the state also needs to be considered in the emergence of an essentially primary urban system. It has been argued that government has played a more important role in urban development in Australia than in any other industrial nation (Logan, Whitelaw and McKay, 1981). All six of the state capitals were partly the result of a highly centralized system of colonial administration. By the second half of the nineteenth century such was the level of support by government for a range of urban functions that Butlin (1959) has referred to it as a form of "colonial socialism." Government support was first given to infrastructure but soon assumed a supporting role to private enterprise via such measures as tariff protection and subsidies. The federal political system which was established in 1901 could be regarded as a logical expression of the settlement system, dominated by six port cities.

The growth of manufacturing, particularly after 1945, was also heavily influenced by government action, and this in turn intensified the dominance of the metropolitan areas. In the 1930s some 16 percent of Australia's GDP was accounted for by manufacturing, and 21 percent of the labour force was employed in this sector. By 1950 the comparable figures were 28 percent and 25 percent respectively. Much of this transformation resulted from Australia's ability to take advantage of international conditions. Demand for a whole range of industrial products far outstripped the capacity of the existing industrial system. Many new technologies and methods had been developed during the Second World War, including methods for co-ordinating large-scale economic enterprises. While some manufacturing development in Australia was homegrown, the rise of the American-based multinational corporation was extremely important. Linge (1967) has shown how state government incentives were extremely important in attracting industries to particular locations. During the 1960s, the Victorian government was extremely successful in bringing factories to Melbourne, perhaps marking the first stage in the divergence of structure between Melbourne and Sydney.

One aspect of government policy which had a profound effect upon both the pattern of industrial development and the distribution of employment was the encouragement of mass migration. Between 1947 and 1976 the nation's population increased from 7.6 million to 13.9 million, the highest growth rate recorded in any industrial nation during this period. Some 60 percent of this increase was contributed by immigration. This population growth increased the size of the home market, provided a labour force for developing industries, and was predominantly located in the large urban areas.

8.3 CHANGES IN EMPLOYMENT

Linge (1979) has demonstrated that by 1966 manufacturing industry had reached a high point, and the modern period of structural change was established. I now turn to a consideration of employment trends from national and regional perspectives.

National trends

At the national level the salient features of the evolving geography of employment may be readily summarized. Thus, it has already been noted that since 1966 the percentage of the labour force in manufacturing industry has declined quite sharply. There have also been important changes in the occupational structure of the labour force. In common with many other industrialized nations there has been a marked increase in the number of people employed in professional/technical and managerial jobs, the so-called knowledge occupations. At the same time, there has also been a rapid growth in poorly paid services activities and clerical jobs. Not withstanding this rapidly changing occupational mix employment remains concentrated in large private and public organizations. In fact, large companies have always been a feature of the Australian economy. Indeed, Connell (1977) has argued that the economy history of the country can be summarized by the nature of the dominant companies in any era. Most banks were formed in the nineteenth century, the industrialization and urbanization of the first half of the twentieth century were reflected in the formation of the large manufacturing and retailing companies, while the resources boom especially after 1960 created the mining conglomerates. Except in the banking and retail industries this high level of economic concentration is also associated with a high degree of foreign ownership which has the effect of truncating the occupational structure to the extent that the top jobs are located elsewhere.

Parallel with corporate power, trade unions have long played an important role in Australia. Indeed, because of union strength it has been suggested that regional wage differences are minimized. While I wish to argue that more research is needed on this idea, it is true that part of the Australian image comprises a notion of egalitarianism and relative equality of wages. There is, however, a growing body of evidence to suggest that income differences among labour market segments are increasing and there are particularly important stratifications based on sex and ethnic origin. It has also been shown that levels of social mobility are not as high as was commonly believed.

Of course, these institutional considerations need to be placed in the context of a rapidly growing and changing population. In this regard, the immigration program which started at the end of the 1940s has continued, albeit not without some important changes in the administration of the scheme. In particular, the "economic" component of migration has been drastically reduced in favour of refugees and family reunions. This has some important implications for the skill levels and employment chances of the migrants. The regional pattern of employment is in part a reflection of these important national trends, but there are significant local effects, and it is to these that the discussion now turns.

Trends in the regional structure of the labour force

The major trends in the distribution of the Australian labour force can be summarized in terms of the changing balance between rural and urban regions, the movement of population to the north and the west of the country, and the changing fortunes of the major metropolitan areas. Each of these trends illustrates the spatial impact of the aggregate changes taking places in labour force, partly in response to external pressures, and partly because of some significant local or regional effects. One aim of this section is to evaluate, however tentatively, the relative weight of factors emanating from the international, national or local levels.

A useful starting point for examining all aspects of these trends is a series of shift-share analyses comparing regional occupational changes in each inter-censual period since 1966 (Maher and McKay, 1986). Between 1966 and 1971 both Sydney and Melbourne gained from the concentration of rapidly growing activities, notably professional/ technical, managerial and clerical occupations, (the industry mix effect) which more than offset a continued drift of population away from these southeast areas to those of the north and west (the regional effect) (McKay and Whitelaw, 1980; Logan and McKay, 1978). Between 1971 and 1976 while both Melbourne and Sydney still enjoyed favourable mix effects, the regional effect became larger, resulting in a total growth rate below the national average. In the latest period for which information is available (1976-81) this pattern has intensified, particularly in the case of Melbourne, where a positive mix effect of more than 7,000 jobs was overshadowed by a regional effect of almost 47,000. In contrast, the smaller capital cities of Brisbane, Perth and Canberra grew more rapidly than the national average throughout the period since 1966, as a result of favourable mix and regional effects. Similarly, large regional effects resulted in gains in most of the rural regions of the eastern coastal zone and in the regions of northern and central Australia.

Turning to the components of these aggregate trends, it is clear that there has been a significant drift in the relative growth rates of urban and rural areas, a population turnaround which has mirrored the "counter-urbanization" tendencies observed in the United States and a number of other countries. Data on internal migration demonstrate that between 1966 and 1971 the state capitals, including Sydney and Melbourne, made net gains from their exchanges with non-metropolitan areas. However, between 1971 and 1976 this pattern was replaced by one of non-metropolitan gains, so much so that by the time of the 1981 census there was an inverse relationship between settlement size and rate of population growth (Jarvie, 1981, 1984; McKay, 1985). As noted, the most significant losses have been experienced by Melbourne and Sydney, but the performance of non-metropolitan areas has been very patchy. Jarvie and Browett (1980) point out that only six of the 43 non-capital city regions did worse in 1971-76 than in the previous period. But even so, the turnaround affected only a relatively small number of regions in any significant way. Table 8.1 shows the details of net migration for each capital city in the last three inter-censual periods. Most obvious is the deterioration in the performance of Melbourne, which had a net loss of more than 55,000 people in the period 1976-81 compared with a net gain of almost 10,000 between 1966 and 1971. Sydney's loss between 1976 and 1981 was even larger, but this was still an improvement on the 1971-76 performance. These two cities are now extremely dependent on migration from overseas to maintain their population totals.

TABLE 8.1

NET INTERNAL MIGRATION BY CAPITAL CITY

	1966-71	1971-76	1976-81
Sydney	-10323	-70802	-57894
Melbourne	9836	-30765	-55507
Brisbane	17532	37478	28548
Adelaide	-1564	14232	-3667
Perth	27713	32975	18554
Hobart	322	2572	-2440
Darwin	5422	183	7533

Source: Maher and McKay, 1985, p. 28

Several hypotheses have been put forward to account for this new situation. Some commentators have argued that most of the loss from the largest cities is caused simply by the spilling over of population from these cities to areas that are beyond their official boundaries. While this may be true in some cases, Maher (1984) has shown that if anything the turnaround has been more complete than had been suspected earlier, with even the smaller metropolitan areas such as Adelaide, Hobart and Canberra demonstrating a declining ability to attract population. Jarvie (1983) has shown that between 1971 and 1976 there was a general reduction in out-migration from country regions, due largely to an increased retention rate of the 15-19 age group. At the same time there was a significant movement out of the largest cities by those aged 25-29 and 30-34, usually accompanied by their children. Hugo and Smailes (1985) have put forward a theoretical framework to account for this change. They suggest that pressures for structural change, many of them originating outside Australia, have reduced the attractiveness, in terms of employment, of the metropolitan areas, which formerly offered many jobs in manufacturing industry. Service sector jobs, which now give employment to the majority of Australians, are less concentrated spatially, and so allow people a greater choice of locations. A growing number seem to prefer the environments and lifestyles offered by smaller towns and rural areas and are able to find employment there.

Many of the non-metropolitan locations which have been growing most rapidly are in the north and west of the country, and this represents an intensification of a pattern of movement which began in the 1860s. Table 8.2 shows rates of inter-state migration by members of the labour force in the periods 1966-71 and 1976-81. These data support the contention by Cordey-Hayes and Gleave (1974) that in industrial economies it is usual for in- and out-migration to be correlated: areas with high levels of arrivals also suffer high rates of loss. Areas experiencing economic problems, and therefore net losses of population, do not have unduly high levels of out-migration. Rather, their loss results from an inability to attract new residents to replace those who are leaving.

TABLE 8.2

RATES OF INTER-STATE LABOUR MIGRATION
1966-71 AND 1976-81
(PER THOUSAND IN LABOUR FORCE)

	Rate in		Rate out	
	1966-71	1976-81	1966-71	1976-81
New South Wales	35	39	39	42
Victoria	34	34	44	49
Queensland	65	89	50	52
South Australia	40	5	54	64
Western Australia	72	59	47	49
Tasmania	53	57	78	75
Northern Territory	398	333	268	234
A.C.T.	335	209	183	201

Sources: McKay & Whitelaw, 1978; McKay, 1984.

The pattern of net migration, and even the total rates of mobility, mask some important variations in rates and patterns of mobility by different occupational groups. Rates of movement by those employed in professional and technical occupations are more than twice those for most other civilian occupations. For this most mobile section of the labour force, exchanges between states have been much closer to balance than for less mobile occupations. Intense patterns of circulation characterize the professional and technical categories, while process worker and labourer categories have been mainly represented in those exchanges where major net changes have taken place, notably in the large loss from Victoria to Queensland. Similarly, the flows to the "frontier" areas such as Darwin and the mining areas of Western Australia also contain a large proportion of process workers/labourers.

These major variations in migration rates and patterns for different occupational groups suggest that a segmented process of migration is in operation, resulting in a long-term change in the regional structure of the labour force. It can also be suggested that different parts of the labour market are subject to quite different dynamics. In this regard, basic differences exist between those organizations concentrated in a few large, often metropolitan locations (what McKay and Whitelaw, 1977, call scale and contact dependent organizations); organizations serving final markets and located in a very large number of locations (market dependent organizations); and organizations located close to particular mineral or energy sources (resource dependent organizations). For those workers moving within such organizations the opportunity for relocation and the spatial pattern of moves will vary greatly between different types of organization (McKay, 1985).

Second, labour markets are segmented according to the place of the organization in the industrial structure. In particular, core organizations have quite different labour structures, career paths and industrial relations policies from those found in peripheral firms. Third, labour markets are segmented according to primary and secondary workers and between "good jobs" or those that are less attractive. Moreover, different labour market segments have been influenced in different ways by changes in the world system.

One obvious set of impacts is the severe pressure being felt by the scale and contact dependent organizations within the manufacturing sector which have faced strong competition from overseas and which have seen the introduction of labour-saving technology. There is some evidence to show that many of the large companies in this sector have tended, as part of a process of rationalization, to contract back to locations at or close to their head offices, predominantly in Melbourne and Sydney. At the same time, there has been some growth of smaller firms in more peripheral locations. Market dependent organizations, with locational patterns similar to those of the total population, have expanded. As the distribution of population has changed, so the operating sites of such private and public organizations have been adjusted. Those that serve large companies and hence cluster around scale and contract dependent organizations have tended to favour metropolitan locations. The automobile industry, and its growing concentration in the Melbourne area, is a case in point. Those small businesses that need direct access to the public have naturally responded to the drift of population to the north and west, and the relocation of small businesses from Victoria and New South Wales to Queensland and Western Australia has been an important component of the migration patterns already described.

Clearly, structural changes in the economy have important consequences for migration. Thus professional/technical workers who are employed in scale and contact dependent activities circulate rapidly between the largest cities in the settlement system. On the other hand, professional employees in market dependent activities circulate rapidly between a large number of metropolitan and regional locations. The area of circulation varies according to the sphere of operations of the organization; some, such as the state education departments, are confined to a single state while others are national in scope. Clerical staff tend to follow the trends operating in the professional/technical categories. However, their skills are more interchangeable between sectors than is the case for professional technical works and spatial mobility is considerably lower. In the case of process workers/labourers, employment opportunities have declined in the traditional manufacturing regions but are available in a wider range of locations. Manufacturing industry, especially in the small firm sector, has decentralized to some extent and some jobs have been created in new mining locations. On the other hand, not all these workers are able to migrate to these new opportunities, preferring to stay in their old communities even in the face of high rates of unemployment.

This brings us to the third theme of this section which is the changing fortunes of the major metropolitan areas, specifically Sydney and Melbourne, as Australia's leading employment centres. The process of the internationalization of capital, or what some have called the transition from monopoly capitalism to global capitalism, and the resultant rationalizations and policy adjustments have affected these two cities in complex ways. Both cities have been affected by the internationalization of investment and financial markets, the emergence of global production, the use of specialized services, and the growing trend towards deregulation of capital and labour markets. Daly (1982; 1984) has shown, however, that the emergence of international capital markets has favoured the growth of Sydney as the dominant financial centre in Australia, rather than Melbourne which had held that position since the mid-nineteenth century. In particular, Daly has argued that the development of a number of major resource developments in the 1960s called for a more sophisticated capital market in Australia and it was via Sydney that the majority of these projects were financed. The growth of trading banks, merchant banks, insurance houses and more recently the licensing of a number of foreign banks in Australia have all favoured Sydney, the site of the country's main international airport and a city with an attractive climate and natural setting. As a result, Sydney's main business area is dominated by financial activities and related services. Melbourne has participated in the growth of some of these activities, but to a far lesser extent.

Sydney has also emerged as Australia's most attractive property market at a time when Asian investment in particular has been growing rapidly. Property markets have become global in nature, in just the same fashion as capital markets. Adrian and Stimson (1986) have argued that Australia's image as a stable and secure long-term investment have encouraged Japanese interest and more recently Chinese property companies. While all capital cities have received some of this investment, Sydney has been most favoured, partly because of the policies of the New South Wales government.

Melbourne, on the other hand, has strengthened its position as a manufacturing centre, and foreign investment has once again been important. While Japanese, Chinese and Korean financial activities and property investors have favoured Sydney, Japanese manufacturing companies have regarded Melbourne as a more attractive location. Melbourne is the long established centre for the majority of Australia's large manufacturing companies and increasingly for a number of smaller, high technology companies. Current policies of the Victorian government are to strengthen this manufacturing base by improving port facilities, already the best in Australia, upgrading the Melbourne airport and improving road links within the metropolitan area. It has been suggested that this manufacturing base is less volatile than Sydney's financial activities, and possibly less reliant on foreign investment. In any event, it seems clear that since the 1960s Melbourne and Sydney have followed quite different development paths, with important consequences for the geography of employment.

8.4 THE CHALLENGE OF THE 1990s: OPTIONS AND REGIONAL IMPLICATIONS

According to Grant, a major potential challenge to Australia's collective psyche concerns if and when some Asian countries follow Japan in surpassing Australia's level of income per capita.

> Australians have begun to realize that their sense of a threat from Asia has been only dimly perceived. The threat of the Asiatic hordes, product of a colonial imagination, has become a challenge of Asian nation-states, offering Australia an as yet undefined, and possibly hazardous, place among them. What is the place and how can Australia gain and hold it? (Grant, 1983, p. 19)

Such questions have produced much soul-searching (see, for example, Scott, 1985), comparisons with Argentina's slide into chaos (Duncan and Fogarty, 1984) as well as exhortations to follow the German, Swedish, Japanese or some other model. These questions also have important implications for regional balance within Australia.

A major factor in Australia's current balance of payments problems has been the decline in the world price for the country's traditional mineral and agricultural exports. The level of indebtedness in rural Australia is very high with predictions of massive farm failures. Decline in the price and volume of coal and iron exports is causing similar problems in particular regions. A number of studies have suggested that demand for agricultural products in the Asia-Pacific region should ensure a good future for Australian agriculture (Bureau of Agricultural Economics, 1984) but some major readjustments may be necessary. Quality controls, marketing and transport improvements are needed and it is far from clear if the traditional products, notably wheat and sugar, are likely to be in demand. Some writers have suggested that it is the wetter tropical areas of the north that are more suited to the type of production that is needed. Even though likely forms of agriculture will be highly capital intensive, and usually controlled by large corporations rather than individual farmers (Laurence, 1987), the regional consequences may be dramatic, especially for the employment base of small regional centres.

The future of the manufacturing sector is no less problematic. The stated policy intention is to move from a highly protected, import substitution form of industry to a more export-oriented range of products. Compared with just about every other industrial nation, Australian levels of protection are high, the effective subsidy from the taxpayers being more than $4,000 million (Anderson and Garnaut, 1987). However, it is proving to be extremely difficult to reduce these levels of assistance to industry because of the geographical distribution of highly protected industries and the likely electoral consequences of any major changes. Clearly, Victoria, New South Wales and South Australia have the most to lose in this process. The depreciation of the Australian dollar, especially in the period between December 1984 and June 1986 when there was a decline of 18 percent against the United States dollar and 45 percent against the Japanese yen, should have had a major impact on the competitiveness of Australian exports, and this has been hailed as the hope of the manufacturing sector. However, a study by the Bureau of Industry Economics (1986) indicated only a modest increase in manufacturing investment.

This lack of investment response has been blamed by the trade unions on a lack of enterprise among employers and has led to calls for the establishment of a national investment fund. Similarly, employers have been blamed for a lack of interest in skill enhancement, training and retraining, all of which are seen as essential to the process of restructuring (Australian Council of Trade Unionism, 1987). The employers have responded that investment cannot be justified in an environment of high wages, restrictive practices and other forms of labour market rigidities. It is perhaps in this area of industrial relations and industrial policy that Australia faces its most serious dilemma. At a time when many writers are heralding the end of "organized capitalism," and especially the high level of union/employment of government co-operation that has characterized many countries (Lash and Urry, 1987), there are still attempts to build such a corporatist approach. In my view the most likely outcome is renewed conflict. On the union side there are serious attempts to amalgamate smaller unions into much larger and more effective units. On the part of the employers, there have been repeated calls for deregulation of labour markets. A number of state governments, particularly that of Queensland, have attempted to remove restrictions on labour markets and have restricted the right to strike in some industries. The most detailed plans for the reconstruction of Australian industries have called for a strong, government-implemented industrial policy to bring some stability and cohesion to what has been a very volatile situation, in contrast to the predictability of single-mindedness of the rapidly growing countries of Asia (Higgott, 1987). Yet in a federal system like that of Australia it is the regional differences, in the form of state governments each with quite distinct political cultures and intent on competing with the other states, that may prove to be the most serious barrier to progress.

9. ADAPTATION TO CHANGE AND UNCERTAINTY: THE SOCIAL IMPLICATIONS FOR AUSTRALIA

D.J. Walmsley

The central thesis of this chapter is that social and economic changes within Australia, coupled with changes in the world order, particularly in the industrial structures of those countries with which Australia interacts economically and politically, are leading to what can be thought of as a "turbulent environment." Exactly how Australia copes with this turbulence will have a major influence on the well-being of the nation and its inhabitants and particularly on whether Australia remains a relatively affluent pluralistic society or becomes transformed into something quite different.

9.1 THE CHALLENGE FACING AUSTRALIA

It is appropriate to begin this chapter by looking at the nature of the changes to which Australia must respond, starting with endogenous changes. In this context four issues stand out as reasonably representative of the changes that are underway: the ageing of the population, the trend towards multiculturalism, changing family structures, and the lack of flexibility in labour relations.

In 1986 there were 1,646,720 Australians (or 10.6 percent of the total population) aged 65 and over. By the turn of the century this figure is likely to have grown to 1,930,000 (11.3 percent of the total population) and by 2021 to 3,494,000 (or 15.9 percent of the population). In the years ahead the fastest growing age group will be those aged 80+ (Department of Immigration and Ethnic Affairs, 1984). These very significant increases in the number of aged people mean that the health care bill for Australia's population is going to increase at a rate greater than population growth. Moreover, much of the increased demand will be for labour-intensive medical care (for example, 24-hour nursing) which is likely to be costly. The trend is significant because it will occur in association with a fundamental shift in the ratio of supporters:supported in the population. Exacerbated by reductions in the retirement age, these trends will have profound implications for the funding of pensions and superannuation benefits. The aged are also likely to want a type of housing stock that is very different from the detached suburban cottage and already retirement villages are mushrooming. Similarly, there seems to be a preference for retirement in resort areas (for example, the Gold Coast, Port Macquarie) that could see these stretches of the coastline turning into a *costa geriatrica.*

Another salient characteristic of the Australian population is its multicultural background. Particularly since 1945 Australia has sustained a massive immigration program that has resulted in a situation where one in five Australians was born overseas. The most prominent groups at the 1986 Census were from: UK/Eire (1,127,000), Italy (262,000), New Zealand (212,000), Yugoslavia (150,000), Greece (138,000), West Germany (115,000). Clearly, then, there has been a European bias to immigration reflecting the "White Australia" policy that endured until the early 1970s. The idea that migrants should be assimilated into an overall "Australian way of life" has gradually given way to a feeling that immigrants should maintain a sense of identity with their cultural origins. Thus the tendency nowadays is to advocate a policy of multiculturalism. Precisely what this means is unclear, although some light was shed on the issue by a number of reports in the late 1970s. These stressed that multiculturalism must promote social cohesion, equality of opportunity, and cultural identity. The reports also noted that the main problem to overcome in promoting multiculturalism is the diffidence and indifference of the host population (Australian Ethnic Affairs Council, 1977; Australian Population & Immigration Council/Australian Ethnic Affairs Council, 1979). This diffidence actually spilled over into acrimonious debate in 1984 when a leading academic put forward the view that the level of Asian immigration (50 percent total in 1984-85 compared to 29 percent in 1982-83) was outstripping public support for such immigration to the point where conflict may result (Blainey, 1984). This view was widely attacked as "racist" but it did have the effect of focussing attention on Australia's immigration selection policies. In the 1970s, the "White Australia" policy had been replaced by a numerical multi-factor assessment system (NUMAS) based on the Canadian model. Basically this sought to assess potential migrants in terms of the contribution they could make to Australia. In addition, Australia has always accepted political refugees. In recent years these have come from southeast Asia. Indeed, between 1975 and 1983 78,000 refugees arrived. Once they have lived in Australia for two years, these refugees are able to take advantage of another aspect of Australia's immigration policy, its emphasis on family reunion. Indeed, it was the relatively large increase in family reunion in the mid-1980s and a concomitant decline in the migration of skilled labour that led to the increase in Asian immigration. The total volume of immigration varies greatly of course: a decade ago it was 53,000, peaking at 107,000 in 1982 before dropping back a little and then increasing once more (128,000 in 1987).

Apart from "displaced persons" at the end of World War II, a good deal of Australia's migrant intake has been in the form of either families or young individuals who married once in Australia. Thus the nuclear family has been a feature of Australia's population profile. In recent years, this characteristic has changed somewhat, particularly with the emergence of single parent families as the fastest growing family type. The reasons are to be found in easier divorce and changing societal values. In addition, the stress of coping with the economic recession that began in the 1970s may have caused the split up of many families. In any event, in the 1986 census there were 277,837 families comprising a single female head plus dependents and 46,334 comprising a male head plus dependents. In the mid-1980s, only 39 percent of female single parents were in paid employment (cf. 84 percent males) (Edwards et al., 1985). As a result a great many families headed by single females are dependent on welfare payments and are therefore on low income. Edwards et al. have suggested that 60 percent of this group have an income level that is below an austere "poverty line." It is not surprising, therefore, that the proportion of children who are the progeny of social service beneficiaries has risen from 4.4 percent in 1973 to 18.2 percent in 1983 (West, 1984). The inability of many supporting mothers to participate in the paid labour force contrasts with "working wives" generally. Out of an Australian labour force of just over seven million, approximately 2.7 million are women and, of these, 1.5 million are married. The percentage of married women who work outside the home rose from 14 percent in 1954 to 40 percent in 1974 (Headlam, 1985). Much of the employment of married women has, of course, been on a part-time basis. The influx of women into the labour force has not therefore had much impact on organized labour, which still tends to be characterized by lack of flexibility and intransigence in the face of change.

Given a labour force of about seven million, Australia has a bewildering number of trade unions (more than 300). As a result, demarcation disputes are common. Often these are petty but sometimes they are of national significance, as with the dispute over which union should be responsible for the building of new grain- handling facilities at Newcastle in the mid-1980s. More commonly, disputes stem from unions seeking to maintain membership levels during times of changing labour demand. Thus the introduction of new express trains in New South Wales led to a dispute over manning levels that brought the state rail system to a standstill in 1983. More worrisome, however, is the way in which current labour relations tend to become institutionalized. Two examples serve to make the point. The centralized wage-fixing system, with state-wide awards, all but abolishes intra-state differentials in wage rates, thus providing little incentive for the mobility of labour or capital. Centralized wage fixing based on the Consumer Price Index (CPI) also works to the disadvantage of those industries (e.g. farming) where commodity prices, and therefore returns, are well below CPI levels. The second example shows institutionalized labour relations at their extreme: the Navigation Act prohibits mining companies from using flag of convenience vessels in coastal shipping, a fact which, according to the Industries Assistance Commission, cost BHP in 1980 an amount equal to 20 percent of the company's steel division's pre-tax profit.

The changes outlined above have been presented implicitly as unidirectional influences on Australian society. In practice, they may not be so predictable. For example, an increase in birth rate, coupled with a large influx of youthful migrants, could dramatically alter Australia's population pyramid. Moreover such changes are not out of the question because the birth rate (births per 1,000 women aged 14-44) has fluctuated in the past from 117 in 1910-12, to 71 in 1932-4, to 112 in 1960-62, to 69 in1980-82 (National Population Inquiry,1975). Likewise there could be a backlash against multiculturalism. Blainey's views could cause some groups to chip away at Australia's bipartisan policy on immigration. Similarly, changing family structure and union intransigence are not immutable facts of life. With respect to labour relations, for example, much has been made of provisions in the Trade Practices Act which allow employers and employees to opt out of the centralized system of award wages and to enter into specific contracts. Such provisions had been the subject of much debate but little action until 1985 when employees at the Mudginberri buffalo abattoir in the Northern Territory agreed on a contract system of work whereby pay was determined on a flat rate for each carton of meat produced (cf. the pre-existing tally system of a minimum workload plus penalty rates).

The timing and extent of such fundamental changes occurring in the nature of Australian society is unclear. Nonetheless, compared to events overseas, these endogenous changes might have implications that are relatively clear-cut. The salient features of the exogenous forces affecting Australia are their unpredictability and uncontrollability. The best known, and most cited, of these events have been the Organization of Petroleum Exporting Countries (OPEC) oil price rises and their contribution to economic recession.

The impact of the fourfold increase in OPEC oil prices in 1973-74 has already been extremely well documented and need not be laboured here, save to say that these increases, coupled with subsequent major rises in 1978-79, contributed to both widespread economic recession and inflation in Australia (Kasper et al., 1980). Given Australia's centralized wage indexation policies, inflation led to higher labour costs to industry. It also led to high interest rates. In other words, a situation was created in the mid- to late 1970s where a cost-price squeeze offered little incentive for innovation or structural adjustment in industry. Indeed, it is a paradox that the pressures for structural adjustment are always greatest when such adjustment would be most painful; in boom times there is little change. Admittedly, the OPEC price rises stimulated interest in alternative energy sources such as coal, but given high domestic interest rates, many businesses borrowed overseas the capital necessary to exploit Australia's resources. The result has been a relatively high level of international indebtedness, equivalent to 35 percent of GDP in the late 1980s compared to 8.3 percent in 1979-80. The servicing of such debt will clearly present problems, especially given that Australia's current account deficit is the third worst in the OECD.

Another overseas development affecting Australia is the development of high technology. Just a few years ago the Committee of Inquiry into Technological Change (1980) took the optimistic view that technological innovation would continue to be beneficial to the Australian economy. The Committee therefore urged the rapid adoption of technology, admittedly with a few safeguards for displaced workers. With the passage of time, attitudes have become less sanguine to the point where high tech is seen as an inevitable but not necessarily positive trend. The doyen of this point of view is probably the Federal Minister for Science (Jones, 1982) who has argued that we must develop new attitudes to work (e.g. job sharing, early retirement) to cope with the diminishing aggregate demand for labour that will inevitably follow the development of high tech. One particular fear is that high tech will herald increased foreign involvement in the Australian economy as orders, even those of the federal government which has 25 percent of the information technology market, are placed predominantly with overseas firms such as IBM and WANG.

Opposition to foreign control in sections of the Australian economy has been very vocal in recent years—and with good reason. Wheelwright (1980), for example, has shown that foreign control extends to 55 percent of mining, 36 percent of manufacturing, 62 percent of chemicals, coal, rubber, petroleum, and plastic, 100 percent of motor vehicles, 91 percent of oil refining, 78 percent of basic chemicals and 55 percent of transport equipment. Moreover, the control is often at the "top end" of the industry. Thus in Queensland in 1981-82, only 15 percent of firms in the mining industry were foreign controlled but these accounted for 71 percent of fixed capital expenditure (Head, 1984). Predictably, given this state of affairs, there arises the possibility that Australia, as Britton argues in Chapter 18 in a manner analogous to Canadian experience, will become a "technological colony" (Hill and Johnson, 1983).

Worries about transnational corporations (TNCs), new technology, and even OPEC, have probably been of less concern to Australians over the years than concerns with overseas trade. This attitude is natural in a country heavily reliant on export markets. In agricultural commodities, Australia seems to suffer at the hands of both the Europeans and the Americans. The Bureau of Agricultural and Resource Economics, for example, has estimated that the European Economic Community's Common Agricultural Policy has depressed world agricultural prices by an average of 16 percent in recent years, costing Australia an average of A$1,000 million a year in the mid-1980s. The US response has exacerbated the position. As a result, times are bad for some Australian primary producers. International sugar prices fell from 24c/lb (US) in 1981 to 3c/lb (US) in 1985, threatening an industry employing 20,000 in small towns on the Queensland coast. Wheat prices also moved down. What is even more worrying is the fact that the world wheat stock in 1986 was 190 million tonnes, equivalent to 12 years Australian production.

The net effect of these exogenous changes, coupled with the endogenous ones, is to create a very volatile and unpredictable environment within which the Australian economy has to operate. Indeed, it seems that the degree and scale of uncertainty or "turbulence" facing Australia is increasing.

9.2 TURBULENCE

The Australian response to increased endogenous and especially exogenous uncertainty has generally been simplistic. For example, some commentators have sought to explain trends within the country as part of the gradual evolution of society from pre-industrial to post-service status (Jones, 1982). Alternatively, much has been made of the need for "structural adjustment," with little appreciation of what that term might mean. Rather than talking about post-service society or structural adjustment, it may be better to recognize that the international environment within which Australia operates is, in many ways, very similar to the concept of a turbulent environment developed by organization theorists (Emery and Trist, 1973; Emery et al., 1974).

Perhaps the overriding feature of turbulent environments is the very rapid rate of change and concomitant uncertainty as to appropriate policies and behaviour. Consider, for instance, how changing demand, overproduction, and intense competition have led to volatile commodity markets: an increase in oil prices led to a search for energy alternatives and, ultimately, overproduction. A second characteristic of turbulent environments is the growth of organizations. This characteristic is particularly marked in the public arena where government monopoly of information encourages the mushrooming of public service bureaucracy, such growth being facilitated by the increased speed with which information and communication can be handled. In Australia, the total number of government employees rose from 824,000 in 1961 to 1,549,000 in 1981, an increase of 88 percent, when the population only grew by 41 percent. Alongside the growth of bureaucracy there has also occurred an increasing emphasis on research and inquiry. At times the generation of yet more data in the form of commissions of inquiry has appeared to be a substitute for action (Walmsley, 1980). Sometimes such an inquiry can have a devastating effect on policy. The National Population Inquiry (1975), for example, exploded the pre-existing myth that Australia would have a turn of the century population of 25 million. It therefore highlighted the inappropriateness of the new cities program and accounts to some extent for the current oversupply of doctors, teachers, and lawyers, the educational infrastructure having been set up at a time of exaggerated population predictions.

The final characteristic of turbulent environments relates to the deepening interdependence of economic and non-economic aspects of life. Government action, and that of major private sector firms, has far-ranging but unplanned consequences. Examples are easily found: the southeast Asian migrant intake may create social unrest while the launching of a satellite (in no small measure because the Canadians had one) is presenting as yet unresolved questions about the ownership and networking of television stations. In these cases, a divergence appears between private benefits and public costs, with governments left with the task of remedying resultant problems, a situation, by the way, which legitimizes further growth in the bureaucracy. Characteristics such as these alter the ground rules for survival and success in society. Given that governments cannot predict what a future state of affairs will be, it becomes impossible to plan an optimal strategy for coping with that situation. Instead the emphasis should be on developing an ability to adapt to events as they unfold. Otherwise inequality and inequity within Australia may increase.

9.3 INEQUALITY AND INEQUITY

Within the context of an increasingly turbulent international environment levels of inequality and inequity have worsened. Trends in unemployment and housing are illustrative. In 1970 unemployment stood at less than two percent but by 1978 unemployment had become a fact of life for much of the teenage labour force and was higher for women and for overseas born. By 1983 the position had deteriorated even more, although it has improved somewhat since. Not only did the economic recession influence the number of people affected by unemployment, it also increased the period of time for which unemployment lasted. It was in this context that the federal government commissioned the Kirby Report and accepted its recommendation of creating 10,000 traineeships by June 1986, with a further 60,000 to follow by 1988. Some authorities have questioned, however, the implicit faith in educational qualifications and job skills that underlies the traineeship scheme and have pointed out that job growth in Australia is at the lower end of the job market. This observation is consistent with trends between 1971 and 1981 which saw increases of 54 percent in junior male labourers, 44 percent in storemen, 278 percent in fast food, 77 percent in waitresses, and 59 percent in shop assistants (Windschuttle, 1984). Such trends, however, will do little to help prevent inequalities and the gulf between the haves and the have-nots.

Inequality in the labour market is parallelled by inequality in the housing market where there seems to be a growing polarization between those who are well-housed, in work, and usually owner-occupiers and those badly or expensively housed, often with incomes in the form of benefits (Paris, 1985). The latter are usually renters. At the time of the 1981 census, 70 percent of individuals in private dwellings were owner-occupiers or purchasers. A further 5 percent were public housing tenants and 21 percent tenants of private landlords. As inflation has pushed up land and house prices, and as unemployment has put home ownership beyond the reach of many, so tenancy has become the only viable option for large numbers of Australians. Since the supply of public housing is limited, many households have been forced to turn to private landlords. Yet in many cities, private rented accommodation is in short supply, a fact which has forced up rents. As a result, many private tenants are paying rents that force them into poverty. Thus, 54 percent of single parents in privately rented accommodation in Sydney in 1984 were in poverty (Henderson and Hough, 1984).

Whether or not this sort of inequality is inequitable is very much a matter of perception. Nevertheless there are certain groups in society which seem to be faring badly and which might therefore reasonably argue that they are being treated unjustly. Women, migrants and farmers constitute three important examples.

In the specific case of women, job opportunities for the 15-19 age group declined between 1971 and 1981, while there was a substantial increase in jobs for those aged 25-44. Indeed the percentage growth in overall female employment during the 1970s outstripped that for males. The picture is not entirely rosy, however. To begin with, there were marked geographical variations in female job opportunities. In contrast to experience in British Columbia, described by Ley and Mills (Chapter 7), the most rapid increases have been in non-metropolitan areas. Some of the increase in rural areas may well have been forced upon women as a way of providing off-farm income to help family properties survive drought and depression.

In addition to geographical variability there are other peculiarities of the female labour market that work to the disadvantage of potential employees. One of these is that most female employment is part-time. Indeed 70 percent of the growth in female employment 1972-77 was on a part-time basis (Power, 1980). Additionally, there is still a great deal of gender segregation in the job market. Given that many "traditional" female jobs have lower wage rates, and that women are over-represented in the ranks of part-time workers, it is not surprising that the ratio of female to male average annual income in 1981-82 was only 0.48, although this was up from 0.35 in 1968-69 (Cass, 1985). This ratio also takes stock of the fact that a lot of women get their income from social security benefits. In fact, in 1981-82, 45 percent of women had such benefits as their main source of income, compared with 18 percent of males (Cass, 1985). Given this lower level of income, it is obvious that female household heads are more likely than males to be tenants rather than home-owners.

The impact of turbulence in the international environment on migrant employment in Australia is also complex. The widely held view that overseas born workers will have to bear a disproportionate share of the adjustment costs associated with structural change at the industry level may not be realistic. Although some groups of migrants are highly sectorally concentrated, for example 48 percent of Yugoslav males were in manufacturing in 1971, and in general migrants earn as little as 87 percent of the average weekly earnings of Australian-born workers (Storer, 1980), their employment prospects may not be significantly different from the rest of the population (Cook and Dixon, 1982). Admittedly, immigrant workers in the clothing, textiles, and footwear industries will continue to experience deteriorating labour market conditions. Yet what seems to be likely is that the success of migrants in dealing with change will be influenced by their fluency in English and their period of residency, as is suggested by the fact that post-1982 immigrants have an unemployment rate of 25 percent, well above that for the rest of the work force (Sloan and Kriegler, 1984). What seems likely is that new technology will encourage jobless growth (that is, increased output without increased labour), the costs of which will be borne by newcomers to the labour force (that is, future migrants and school leavers) rather than by those presently in employment.

In the case of farmers the threat to incomes and employment comes not from new technology but from a cost-price squeeze brought on by a fall in commodity prices and an increase in production costs. One third of farmers surveyed by the Bureau of Agricultural Economics in 1985 had negative incomes and only survived by supplementation from off-farm income. On top of this, of course, computerized livestock and wool auctioning is causing a centralization of services traditionally provided by country towns, causing many areas to suffer population decline (Rural Development Centre, 1985).

The plight of much of the rural sector, and of many country towns, raises the question of whether or not certain *regions* are faring particularly badly in terms of economic and social development. The western suburbs of Sydney and Melbourne are commonly cited as problem areas. So too are inner city areas. For the most part, however, debate about spatial variations in well-being has focussed on inter-state comparisons. This debate has led to what has become known as the "national fragmentation thesis." According to this proposition, shifts in regional economic power resulting from the mineral resources boom have led to a fragmentation of the national economy and exacerbated political strains in the Australian federal system. In particular, Western Australia and Queensland, the main beneficiaries of the resource boom, have extended their power at the expense of the other states. Thus, according to this thesis, there has been a vertical shift of power away from Canberra and a horizontal shift of power from New South Wales and Victoria to Western Australia and Queensland (Head, 1984). The thesis is a simple one and one with a certain appeal. For example, several publications have recently looked at the inappropriateness of the federal system in coping with the sort of economic change that seems to be occurring in Australia (Aldred and Wilkes, 1983; Patience and Scott, 1983). The evidence in support of the thesis is equivocal, however, because much of the benefit from resource investment leaks back to head offices in Sydney and Melbourne and even state-based royalty revenue may be offset by infrastructure costs.

9.4 MALADAPTIVE AND ADAPTIVE STRATEGIES

The extent to which differences in well-being between one part of the country and another become significant will be determined in large part by the activities of Australia's various governments. In this respect it is appropriate to consider what have been identified as maladaptive and adaptive strategies in coping with turbulence (Emery et al., 1974). Such strategies are set out in Table 9.1. The point I wish to make is that the two sorts of strategy are really opposite ends of a spectrum. Moreover, identification of appropriate strategies from this table is no touchstone to success because it is not unknown for adaptive strategies to be applied ineffectively.

TABLE 9.1

MALADAPTIVE AND ADAPTIVE STRATEGIES

Maladaptive	Adaptive
CENTRALIZATION OF CONTROL & DECISION MAKING. Seeks to simplify turbulence to a matter of scale and fallaciously equates size with efficiency	DECENTRALIZATION OF CONTROL. Encourages organizations to be flexible and able to respond quickly to local conditions
SPECIALIZATION AND SEGMENTATION. Divides problems into smaller, manageable ones thereby ignoring interrelatedness and reducing the scope of policy choice	DECREASED SPECIALIZATION. Encourages holistic perspective that reduces chance of organizations pursuing ends without respect to overall well-being
DISSOCIATION OF PEOPLE FROM ISSUES. Seeks to reduce turbulence by ignoring problems. Problem cases often dismissed as deviant	PUBLIC PARTICIPATION. Encourages continuous education so that individuals can understand and participate in social policy
SUPERFICIALITY. Ignores fundamental issues and looks instead at surrogate issues that are easily quantified, responding first to this, then that, in a way unrelated to the fundamental issue	OPEN PLANNING. Strives for open debate about fundamental social concerns

Source: After Walmsley (1980)

Several maladaptive strategies are currently in vogue in Australia. The first is the emergence of 'Big Government'. This strategy is maladaptive because it involves not only growth of bureaucracy but also the concentration of decision-making power in one or two key departments (for example, the Treasury, the Prime Minister's Office). Thus there can arise a situation where, because of centralization, governments become too big and unwieldy. A second maladaptive strategy is the compartmentalization of welfare issues into different domains of responsibility. Thus "youth" and the "aged" now have their own "offices" in some states and at federal levels. The danger with this is that common problems may not be fully appreciated. Compartmentalization of welfare responsibility in Australia is, of course, institutionalized under S.51 of the Constitution, whereby the federal government is responsible for cash payments of pensions and benefits (income security) while the states have responsibility for social services (health, education, housing) (Walmsley, 1984). This situation makes social planning difficult. However, it is exacerbated by the fact that Australia has, by international standards, a very high level of vertical financial imbalance with the federal government having far more resources than commitments and the states more commitments than resources. Thus, the states' ability to provide services in their area of responsibility (for example welfare) is very much contingent on the trickle down of funds from a higher level of government.

A third maladaptive strategy involves a preoccupation with the symptom rather than the causes of problems. In particular there is a tendency to reduce welfare considerations to monetary issues and to look mainly at issues that are easily measured (e.g. average weekly earnings) or at least quantifiable (e.g. poverty line) rather than at important but vague concepts (e.g.. well-being). And, of course, public attention to these figures gives them validity no matter how inadequate they might be. One spinoff from a preoccupation with quantification is an attitude of mind that brands people who are different from the mass as in some way "deviant." Thus the unemployed can become labelled "dole-bludgers." Such labelling is maladaptive because it seeks to oversimplify complex issues. Unfortunately such simplication also extends to Australia's view of its place in the world. For example, southeast Asia is often thought of as a uniform region that gains its pre-eminence in export markets simply as a result of low wage rates. There is little widespread appreciation of Japan's sophisticated economic planning or of the current slowdown of growth rates in east Asia. The prevailing attitude seems to be that if the Americans and Europeans would only play fair, if the Arabs would only behave, if China develops quickly, and if the Third World follows Western consumption patterns for beef and sugar, then all will be well.

Tariff protection constitutes a fourth maladaptive strategy because such a policy may fail to recognize the shifting environment within which Australian producers operate. To subsidize an ailing industry may be silly if that industry has little or no possibility of recovery. To impose tariffs is also to ignore political realities in the western Pacific. To hope that rapid regional economic growth will lead to import demand which Australia can serve is shortsighted so long as Australia is reluctant to allow access for overseas produced clothing and textiles (Garnaut and Anderson, 1980). Tariff protection is, however, very much a sacred cow. Even the much vaunted plans for the car industry and for clothing, textiles and footwear seek only to reduce protection to 57 percent and 50 percent respectively.

Finally, perhaps the most significant maladaptive strategy for coping with turbulence in Australia is to be seen in the policy of fiscal equalization. At the moment much of the tax-sharing in federal-state financial transfers is based on the idea that all states should be put on an equal tax footing. Thus Tasmania gets twice the per capita amount given to Victoria. Likewise in state to local transfers, equalization grants can be up to two thirds of the total money involved. Such equalization policies are a negation of basic geographical variability in resource endowment and have the unfortunate consequence of fossilising settlement patterns by encouraging an unwillingness to move.

All this discussion of maladaptive strategies may seem rather abstract but there is little point in focussing on specific tactics or programs until the appropriateness of the overall thrust of government policy has been settled. A central argument in this chapter is that this point has not yet been reached. It seems a pity to end on a depressing note but the fact of the matter is that, to date in Australia, there is very little evidence of governments moving towards what could be considered adaptive strategies. There is very little evidence of any move towards decentralization of control in social policy. Thus it is unlikely that Australia will see a situation develop where social policy organizations can respond quickly to local conditions. Likewise there seems little public debate about social policy. Perhaps Australians are unconcerned about the future. Unemployment, for example, is not a burning issue because, even allowing for those involved, their dependents, and those with a social conscience, the fact of the matter is that over 80 percent of the population is unaffected. Given this less than comfortable existence, it is perhaps not surprising that so little attention focusses on "social futures."

ACKNOWLEDGMENTS

I would like to thank the Department of Human Geography at the Australian National University for hospitality and a Visiting Fellowship in 1985. Although my visits to Canberra were all too brief, they did enable me to do much of the reading on which this paper is based.

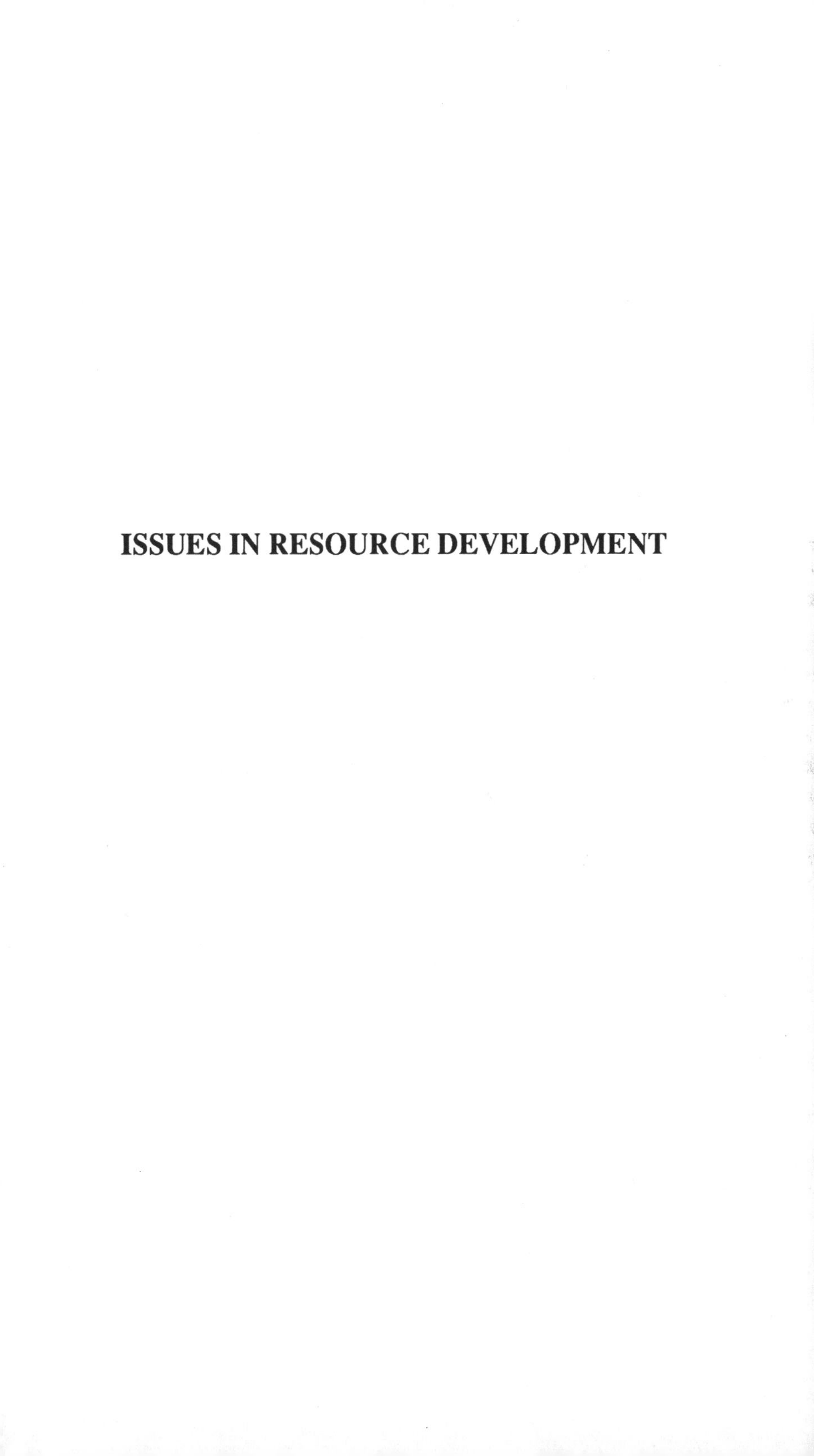

ISSUES IN RESOURCE DEVELOPMENT

10. NATURAL RESOURCES AND ECONOMIC DEVELOPMENT: CANADIAN PERSPECTIVES

Thomas Gunton and John Richards

During the last several decades, there has been considerable debate on the role of natural resources in regional development. Advocates of comparative advantage argue that regions with a plentiful supply of natural resources such as western Canada should concentrate on what they do best, produce natural resources. Dependency theorists on the other hand argue that the production of natural resources can never lay the foundations for a healthy economy. The preferred alternative, they believe, is to develop a diversified manufacturing base. In our opinion pursuit of regional comparative advantage is a more persuasive argument. The realization of regional comparative advantage in natural resources, however, requires an aggressive and well conceived development policy.

This chapter will identify the elements of an effective resource policy for regional development. It begins with a discussion of the role of natural resources or "staples" in economic development, followed by an analysis of the dynamics of policy making in the resource sector with particular reference to western Canada. For the purposes of this chapter "staple" industries are defined as primary activities such as mining, agriculture and fishing and "primary manufacturing" (Bertram, 1967, p. 75). Primary manufacturing includes activities which do not need elaborate processing, such as pulp, in which the value added by manufacturing is low relative to the value of the staple input, or highly capital-intensive activities such as the production of simple petrochemicals in which the staple is upgraded into some standardized intermediate good. In the Canadian context such staple activities are important and are largely exported (Table 10.1).

TABLE 10.1

ESTIMATE OF STAPLE-RELATED EMPLOYMENT, BY SELECTED REGIONS, 1981

	Prairie Provinces	British Columbia	Canada
		Percentage	
Resource[1]	15.9	8.3	8.1
Closely linked manufacturing[2]	2.6	8.4	57
Final demand linkage[3]	18.5	16.7	13.8
Total, staple related	37.0	33.4	7.6
Other	63.0	66.6	72.4
Total	100.0	100.0	100.0

1 Includes agriculture; forestry; fishing and trapping; mines (including milling), quarries and oil wells; electrical-driven, gas and water utilities.

2 Includes the following manufacturing activities; food, beverage (including food processing, flour milling) wood (including sawmills); paper and allied products; smelting and refining: petroleum and fuel products (including refining).

3 The indirect staple employment arises from a multiplier of 2 applied to direct employment. This multiplier value is a conservative estimate drawn from a survey of the literature.

Source: Calculated from *Census of Canada*, 1981.

10.1 STAPLES, RENT AND GROWTH

As a point of reference let us sketch a simple staple model of regional economic growth. Assume a region possesses an abundant low cost supply of some natural resource and the staple produced from it enjoys strong international demand. Further, assume economies of scale are important in manufacturing activities and the region faces high transport costs to major markets. What are the implications for economic development? The high transport costs provide some protection against manufactured imports and enable some local manufacturing to emerge, but simultaneously they deny the ability of local manufacturers to compete in export markets. Restricted to the local market they cannot realize the scale economies of manufacturers located near large concentrations of population. Conversely, despite regional staple producers having to absorb transport costs to export markets, the comparative advantage in a staple industry may permit not only realization of normal returns to labour and capital so employed, but also economic rent. To maximize regional per capita income the region should specialize in staple exports, importing its manufactured requirements.

The addition of "linkages" and "leakages" affords a rudimentary dynamic perspective (see Watkins, 1963). As income from the staple industry is spent locally the supply of locally produced goods and services for household consumption ("final demand linkages") is increased. This type of multiplier effect is limited by "leakages" out of the regional economy in the form of expenditures on imported goods, savings, and taxes levied by the federal government. Over time new staple industries may also be developed which generate "backward" and "forward" linkages. The former refer to income created in industries that expand to provide inputs to the staple sector; the latter to income in industries which process the staple. The regional economy may then diversify into secondary manufacturing, via one of the three linkage mechanisms.

An early advocate of "staples as catalyst for growth'" is W.A. Mackintosh. In his classic economic history of Canada prepared for the Rowell-Sirois Commission (Royal Commission on Dominion-Provincial Relations), he was adamant that "rapid progress in...new countries is dependent upon the discovery and development of cheap supplies of raw materials by the export of which to the markets of the world the new country may purchase the products which it cannot produce economically at that stage of its development" (Mackintosh, 1964, p. 13). Writing at the end of a decade of economic depression whose most severe impact had been in the agricultural economy of the prairies, no one could be unambiguously sanguine about staple-led growth. Admitting that "incomes derived from the export of raw materials are notoriously variable", Mackintosh summarized the central problems of any staple-based development as "those that occur when fluctuating incomes are coupled with the rigid expenditures occasioned by heavy debt charges" (Mackintosh, 1964, p. 13). According to Mackintosh the debt charges stem from the need of newly developing countries to import capital, technology and equipment. While Mackintosh was acutely conscious of problems with resource-based growth, his emphasis was to improve the quality of public and private resource planning.

The other classic analysis of Canadian staple development is by Harold Innis. For Innis, the staple theory was more than an export-led growth theory. It was a framework for analysing the entire political, social and economic history of Canada. Innis attempted to show how Canadian history has been shaped by the export of a succession of staples from the Canadian hinterland to metropolitan European and American markets. Staple development is not, for Innis, a benign one: "Each staple in its turn left its stamp, and the shift to new staples invariably produced periods of crisis in which adjustments in the old structures were painfully made and a new pattern created in relation to a new staple" (Innis, 1972, pp. 5-6).

According to Innis the imperial state dictated the pattern of staple development in the colony, controlled the terms of trade and captured most of any rent arising. Citizens of the peripheral colony were forced to finance from their taxes much of the fixed overhead (such as railways, canals, roads) required to extract and transport the staple. The economy of the colony thus became "dependent" on continued exports of the staple to the imperial state. As Innis concluded:

> The economic history of Canada has been dominated by the discrepancy between the centre and the margin of western civilization. Energy has been directed toward the exploitation of staple products and the tendency has been cumulative Agriculture, industry, transportation, trade, finance and governmental activities tend to become subordinate to the production of the staple for a more highly specialized manufacturing community. (Innis, 1956, p. 385).

Innis inspired many contemporary critics of staple development such as Watkins (1963, 1973, 1977), Naylor (1975), Drache (1978) and Britton & Gilmour (1978). Watkins and Naylor, for example, illustrate how staples engender an "export mentality" among entrepreneurs who become content to reinvest savings into staple industries and shun diversification into new sectors. This process is reinforced by foreign-owned firms which have a bias for locating linkages outside the staple supplying region, because such linkages are typically already established, frequently within the corporate structure of the firm itself. Governments are hampered in their attempt to implement resource policy because they have a weak bargaining position vis-a-vis large multinationals which control the savings, investment, technology and markets necessary for the functioning of the staple economy. In sum, the externally controlled staple economy can be locked into a staple trap. As Watkins observes:

> At the root of our problem lies foreign ownership. In the resource industries it means the export of staple in unprocessed form and the outward drain of surplus. In the manufacturing industries, it means truncated, branch-plant structure with a high propensity to import parts and a demonstrated incapacity to export finished products, or even to hold the Canadian market without tariff protection. The combination is deadly (Watkins, 1973, p. 115).

Whether or not staple dependence is inevitable is debatable. As Bertram (1967) has shown considerable manufacturing investment has occurred, and the evidence of entrepreneurial resistance to industrial diversification is dubious (Macdonald, 1975). Similarly, Richards and Pratt (1979) illustrate that policy makers in western Canada were successful in the 1970's in gaining control of staple industries and using the rents for regional capital formation. Their analysis also illustrates that the preferred alternative of some neo-Innisians to shun staples and concentrate on domestically owned manufacturing failed. The neo-Innisians are correct in emphasizing the problems of staple-led growth. It should not be concluded, however, that staples should be shunned in favour of manufacturing because of these problems.

Coming from a different perspective, Chambers and Gordon (1966) developed a model of the Canadian wheat boom which measured the contribution of natural resources to economic development. In the simplest version of their model they considered the immigrant leaving the boat in Montreal to have faced two basic options: manufacturing or farming. They assumed all capital to be borrowed from abroad at a fixed price; thus profit and interest income accrued to foreigners. They further assumed that Canada, a small country, could export and import within the relevant range whatever quantities it desired of wheat or manufactured goods, without affecting international prices. The final assumption was constant returns to scale in both farming and manufacturing. Competition implied the equalization of wage rates across sectors as workers migrated to maximize income. In the case of farming, tenant farmers earned the prevailing wage, but the income of homesteaders was analytically divided into two components: the wage income they could earn in alternative employment and rent. Under these assumptions the effect of the immigrant's choice on national income reduces to the size of any agricultural rent.

Chambers and Gordon then posed the hypothetical question, "What would have happened if all the (prairie) land that was brought under cultivation between 1901 and 1911 had been impenetrable rock?" (Chambers and Gordon, 1966, p. 317). The immigrants' occupational choice would have been restricted to that of wage earner. The manufacturing sector could have expanded to absorb all these frustrated farmers at constant prices, labour productivity and wages. Apart from a trivial loss of tariff revenue on manufactured imports displaced by enlarged domestic production, the only effect on per capita Canadian incomes would have been the lost agricultural rent. This they estimated was merely 8.4 percent of the observed per capita income increase of Canadians between 1901 and 1911.

The two theoretical conclusions from this model are: 1) for a "small country" possessing both a staple export sector and a manufacturing sector subject to net imports, labour employed in the staple sector could be employed in the manufacturing sector, and 2) the economic benefit derived from a natural resource endowment is the rent generated; not the aggregate income.

Dales, McManus and Watkins (1967) criticized Chambers and Gordon for defining the impact of staples on intensive growth (output per person) while ignoring its impact on extensive growth (total output). They also pointed out that Chambers and Gordon's small estimate of the contribution of staples to economic growth is based on dubious assumptions in their model. Although his analytic treatment is flawed, Lewis (1975) emphasized perhaps the most serious analytic weakness of the model, namely its trivialization of the demand side analysis. The model adopts the "small country" assumption of international trade theory that Canada faced an infinitely elastic demand at fixed international prices (plus the tariff where relevant) for the commodities it bought and sold. The "small country" assumption is reasonable for wheat, a homogeneous commodity of which many industrial countries of Europe had become net importers. In the absence of prairie land Canada might well have become a net exporter of manufactured goods and importer of grain. But international markets for manufactured goods are less competitive than those for homogeneous primary commodities and the export demand for Canadian manufactures should be assumed of finite elasticity (i.e. downward sloping).

The hypothetical exercise of reducing available prairie land now results in complex shifts in the supply of, and demand for, labour in the two sectors. In general, the number of farmers declines; some displaced farmers are absorbed in manufacturing, but this sector can expand only by a lowering of Canadian manufactured good prices and the prevailing wage. To the extent these goods are produced and consumed within Canada, a decline in their price merely redistributes income from industrial workers to Canadian consumers; to the extent they are exported the benefits of the price decline are shared with the rest of the world.

In summary, the exercise of eliminating prairie land lowers per capita Canadian income levels because of, first, lost agricultural rent; second, a decline in average output per farmer resulting from the wage decline which, in turn, lowers the opportunity cost of homestead labour and induces substitution of labour for land in wheat production; and third a negative terms of trade effect arising from the decline in the price of Canadian manufactured good exports.

Notwithstanding its limitations, the analysis and conclusions of Chambers and Gordon have been enthusiastically taken up by, among others, the Economic Council of Canada. Copithorne (1979), for example, estimated rents in the Canadian resource sector. He concluded that rents are unimportant in most cases and are not therefore a significant factor explaining regional development.

Capithorne is correct in identifying rents as the principal benefit derived from natural resource development. His conclusion that rents are insignificant, however, is based on dubious assumptions. Like Chambers and Gordon, he assumes that labour in the resource sector could be employed in the manufacturing sector at the same wage. He further assumes that the resource sectors are operating at optimum levels of competitive efficiency, and therefore provide little potential to generate rent by improved managerial efficiency on intramarginal production. Finally, he assumes that high elasticity of supply by producers beyond any provincial government's jurisdiction would frustrate attempts to generate monopoly rents by curtailing production to increase price.

A recent study by Gunton and Richards (1987) shows that these assumptions are incorrect. While Copithorne is correct in stating that labour employed in the resource sector could be employed elsewhere in the economy, he is incorrect in stating that it could be employed at the same wage rate. He is also incorrect in assuming that resource industries are efficiently run and that there is little potential to increase rents through the judicious use of market power. As Gunton and Richards show, substantial rent is dissipated by inefficiency in the resource sector and potential rents are foregone by the failure to exploit market power. Correcting for these errors, Gunton and Richards illustrate that resource rents are substantial and that natural resources are consequently a significant factor explaining regional development. The problem is with natural resource policy; not natural resources. This is the issue to which we now turn.

10.2 CANADIAN RESOURCE POLICY

A prominent economic historian (Aitken, 1959) characterized Canadian economic policy as the promotion of growth by subsidized infrastructure and unrestricted access to natural resources. Canadian politicians have never, in his mind, aggressively managed resource development to maximize rents and use them for regional capital formation. Other studies provide considerable evidence suggesting that Aitken was right (Kierans, 1973; Mathias, 1971; Richards and Pratt, 1979; Gunton and Richards, 1987). Indeed, public policy has often reduced economic growth by subsidizing uneconomic resource development, thereby consuming capital which could be used more productively elsewhere in the economy. The increased capacity has also reduced commodity prices of the region's exports.

During the late 1960s and early 1970s there was substantial criticism of the strategy of subsidizing the resource sector. The Carter Commission, in its comprehensive review of the Canadian tax system, called for the elimination of many concessions enjoyed by extractive industries as inequitable and inefficient (Canada, 1966). In Ontario, the Smith Committee used the concept of rent to justify imposition of special royalties on the resource sector (Ontario, 1967). And in his pivotal analysis for the Manitoba government, Kierans argued that if governments did not collect rent, multinational firms would leak it from the province thereby impeding regional economic development and possibly leading to outmigration (Kierans, 1973).

Combined with the commodity boom in the early 1970s, the criticism of subsidizing resource development precipitated significant changes in government policy. Between 1969 and 1975 the government of Manitoba initiated a series of reforms — an innovative royalty system, a crown corporation to undertake exploration and development of new mineral deposits, a significant tightening of leasing regulations and a compulsory joint venture scheme. Although moderate compared to Kierans' proposal for "nationalizing" the industry, cumulatively these changes were important. In Saskatchewan the government pursued similar reforms including creation of a Crown mineral corporation to engage in exploration, tightening of lease requirements, introduction of new royalty systems for rent capture in oil, uranium and their industries and, from 1976 to 1978, takeover of nearly onehalf the assets of the provincial potash industry. A primary motivation for this last initiative was uncertainty over the constitutionality of rent capture by less interventionist means. Between 1972 and 1975 the government of British Columbia implemented a new mineral royalty, launched a major royal commission on reform of forestry management and acquired several major firms in the forestry sector. Although the primary motive for acquisitions was to preserve employment in single industry towns threatened by mill closures, government leaders were also acting to increase public control of the major provincial resource industry. Elected in 1971, the Alberta conservatives became acutely conscious of the magnitude of windfall rents created by post-1973 oil and gas price increases, rewrote their royalty legislation and resorted to the use of Crown corporations in the resource sector. The federal government also showed increasing interest in resource rents. In 1972 Ottawa increased its resource taxation by legislating some of the Carter Commission reforms. As oil prices rose in 1973, the federal government simultaneously acted as agent for oil-consumers by controlling domestic prices and competed with the oil producing provinces to capture the windfall rents. The climax of the federal government's interest in rent collection came in 1980, with the battery of new taxes, new regulations and nationalization of private assets contained in the National Energy Program.

It is rash, however, to conclude from this history the existence of some consistent process whereby provincial and federal politicians learned the intricacies of resource development and improved their policies accordingly. The magnitude of windfall resource rents in the 1970s was sufficient to break the regulatory inertia in all provinces and the federal government, but by the time the new taxes and royalties were in place rents in many resource sectors had disappeared; boom had become bust. Lobbyists within resource industries then blamed — with partial justification — their economic woes on new government initiatives to capture rent. Public support for government rent capture waned as investment, employment and profits diminished in the resource sector.

Such was the fate of British Columbia's 1974 policy of "super royalties" applied to minerals. Public opposition to the province's resource policies contributed to the electoral defeat of the government in the 1975 provincial election. A newly elected Social Credit government eliminated the controversial mineral royalty, sold off the former government's acquisitions in the forest sector and abandoned the process of increasing the public share of forest rents. The government has returned to the policies of the 1960s of subsidizing uneconomic resource megaprojects such as Northeast Coal development and the Revelstoke dam.

A more dramatic example of lags in political response is the federal National Energy Program (NEP). Ottawa undertook this initiative only after the second round of OPEC price increases in 1979, seven years after the first round. By the time NEP initiatives were in place, the 1982 depression had begun and real oil prices were falling. Lobbies within the oil industry blamed the NEP for the industry's frustrated expectations and, particularly in western Canada, succeeded in discrediting among many people the idea of rent capture by the federal government. In 1985, the federal government abandoned the NEP, and via the "Western Accord" transferred substantial energy rents back to producers. In mining, a major federal policy paper concluded there to be little rent in the sector and that attempts to capture it would reduce incentives to invest and distort production decisions by established producers (Canada, Energy Mines and Resources, 1982, p. 54). The Alberta and Saskatchewan governments, meanwhile, have substantially reduced royalties in the oil and gas sector and increased subsidies for resource megaprojects such as the heavy oil upgrade in Saskatchewan and for exploration.

In sum, despite the furor over public rent collection initiatives of the last decade, the proportion of rent, actual or potential, captured by government has been modest (Table 10.2). It appears that public policy has been unsuccessful in maximizing the contribution of natural resources in regional development. We will now utilize lessons from this experience to identify the elements of an effective resource policy.

TABLE 10.2

ESTIMATE OF RENTS IN SELECTED RESOURCE INDUSTRIES

Industry	Annual Rent Collected by Government	Potential Aggregate Annual Rent
	(1985 $m)	
Manitoba Nickel Industry	$ 35	$ 161-224
Manitoba Hydro	$ 11	$ 165
Saskatchewan Potash Industry	$ 6	$ 85-156
British Columbia Forest Industry	$ 488	$ 1032-2157
British Columbia Hydro	$ 184	$ 765-1609
Pacific Fishery	$ 3	$ 83

Source: Gunton and Richards, Resource Rents and Public Policy, 1987, p.21.

10.3 EFFECTIVE RESOURCE POLICY

Rent Maximization

The primary objective of natural resource policy should be to maximize resource rents. One way of doing this is to reduce costs. Regulations which attempt to achieve ill-defined social objectives on resource developers at the cost of foregone rent should be removed. Several recent studies of the British Columbia forest industry, for example, have emphasized the costs of policies such as export restrictions on logs and regulation of cuts (Percy, 1986). Studies of the mining industry discuss costs resulting from exploration obligations and processing requirements (Kierans, 1973). While there is justification to regulate resource industries to realize social goals, pursuit of such goals should be via explicit and accountable public policy not via implicit *quid pro quo*. For example, instead of increasing employment in the BC forest industry by banning log exports, the government could provide direct subsidies for domestic log processing. The costs and benefits of achieving the social goal become more visible and easier to evaluate.

A second way of reducing costs is to prevent development of surplus capacity. This is particularly a problem in the resource sector because of the volatility of commodity prices and the capital intensity of the projects which means that capital cannot be reallocated once invested. Indeed, as a number of case studies on resource industries as diverse as forestry, potash, hydro, and nickel (Gunton and Richards, 1987) illustrate, resource industries have an unfortunate tendency to dissipate rents by excessive expansion during resource booms. While it is impossible to have perfect information regarding future markets, much improvement is possible.

First, governments should remove subsidies on infrastructure which encourage uneconomic expansion. Without the massive subsidies for rail and townsite construction for the British Columbia Northeast Coal project, for example, the private investors would not have proceeded and the private and public sectors would have been spared the massive losses associated with this uneconomic project.

Second, given the large-scale nature of resource projects, efforts should be made to co-ordinate resource investments to prevent individual producers from developing capacity to serve the same market. This propensity for the so-called pre-emptive strike of investors competing for the same market by committing first on the dubious assumption that other producers will then withdraw can be reduced by comprehensive indicative planning undertaken by a joint agency representing private producers and government which is mandated to analyse market conditions and identify the most efficient expansion strategy. Such an agency could have prevented the BC Northeast Coal development project, for example, in favour of expansion in the southeast coal fields. Finally, the persistent "manic" expectations identified by Mackintosh, among others, must be dampened. Uncertainty regarding future resource markets makes some frustrated expectations inevitable but, given past experience, policy makers need to err on the side of pessimism. Possible foregone benefits due to capacity constraints during booms can be substantially less than the potential costs of unused capacity and labour during slumps.

Policies to maximize resource revenue have received less attention than policies to minimize cost, despite growing evidence that prices can be influenced by producers or consumers. Producers such as OPEC have been very successful in increasing petroleum rents by restricting output, while consumers such as the Japanese have been successful in appropriating rents by encouraging surplus capacity and using consortium purchasing to lower commodity prices. To realize market power or to counteract consortium buying, resource producers should be encouraged to form their own cartels. For example, joint private/public marketing boards could be created for major commodities. These boards would undertake the indicative planning to prevent surplus capacity as well as regulate production and negotiate as one supply to realize market power. Gunton and Richards (1987) show that policies to co-ordinate investment and regulate output could have increased natural resources rents in various sectors. Certainly, a marketing board would have increased BC coal producers' bargaining power with the Japanese, especially if suppliers from overseas, for example in Australia, were included in the selling cartel and governments.

Rent Collection

It is commonly held that attempts by governments to collect rents discourage investment or distort production decisions. Therefore, there is an alleged trade off between economic welfare and rent collection. This view is misleading. Rent is a surplus after compensating all costs of production including a "normal" or sufficient return to capital to entice it into production. Collecting this surplus should have no effect on investment or production. Instead, collection of rent will actually improve regional economic welfare by retaining the surplus, which might otherwise be leaked to compensate external owners, as regional income. Further, if rent is not collected it can act as a subsidy which distorts relative prices and hence a firm's production decision. This subsidy allows firms to earn normal returns without due regard for costs and to retain income within the sector, income which could be more productively invested elsewhere in the economy. Uncollected rent can also be passed onto workers in terms of wages. Dissipation of rents in high costs, surplus capacity or higher wages can increase rigidities in the economy which impede diversification or reallocation. In the event of market decline, it is easier for the economy to accept a reduction in government resource revenue than bankruptcy of inefficient resource firms or major wage reductions among resource-based workers.

Case studies of selected resource sectors in western Canada confirm these concerns. Gunton (1987), for example, shows that in the nickel industry of Manitoba, rents were dissipated by building excess capacity and incurring higher than necessary operating costs. Remaining rents were leaked from the regional economy to fund Manitoba's competitors. Overall, the dissipation and leakage of rents resulted in employment being less than onehalf the potential.

Mechanisms for capturing rent are evaluated by Gunton and Richards (1987). Royalties and competitive bidding are the most common approaches. However, the domination of most resource sectors by large multinational firms renders these passive rent collection techniques ineffective. Resource firms are able to circumvent royalty regimes by intracompany transfers of revenue and by using their economic control over investments and production to force government concessions. Competitive bidding is impeded by insufficient competition. Neither of these approaches gives the public the control of private output decisions necessary to maximize rent. Therefore, some degree of public ownership in staple industries is necessary to acquire the data, bargaining power and control over production that is necessary to capture rents (Kierans, 1973; Cohen and Krushinsky, 1976; Richards and Pratt, 1979; Gunton and Richards, 1987). As Richards and Pratt conclude in their study of the western Canada economy:

> A *sine qua non* for a government to capture rent and other benefits is that its leaders be ideologically oriented to take entrepreneurial risks and be prepared to reject a low level of rent available with certainty and bargain for a higher

> level which may prompt prospective industries to retaliate, be prepared to invest public funds if the price of private investment is too high, be prepared to sacrifice immediate for probable but uncertain future benefits (Richards and Pratt, 1979, p. 327).

As Bernard and Payne (1987) show in their study of hydro power, public entrepreneurship is not without its problems. Like their private sector counterparts, public corporations can dissipate rent through excess capacity and high operating costs. Crown corporations may be particularly vulnerable to political demands for current investment and job creation. Clearly, if public entrepreneurship is to be effective, the enterprise must be given explicit instructions to maximize rent and the rent must be paid directly to the owners instead of being retained in the corporation where it could tempt unjustified expenditures.

Rent Distribution

Norrie and Percy (1982) evaluated the impact of alternative rent distribution mechanisms on economic welfare in Alberta. Under hypothetical assumptions they determine the effect arising from a one percent energy price rise. Under the base case ("classic staple boom") private agents own the reserves and capture all incremental rent. In the long run, allowing interregional migration of labour and capital, their model predicts a per capita GDP rise of .11 percent. They then consider three options for provincial government intervention. First, fiscally induced migration, in which government captures resource rent, taxes capital and disburses the revenue via a fiscal dividend to all residents, regardless of length of residence. Second, the expansion of public sector output where government captures resource rent, taxes capital and uses this revenue to finance expansion of public sector output. The third option is the same as the second except that the government also insulates provincial primary manufacturing sector from the energy price rise. Under these options a one percent energy price rise induces, in the long run, changes in Alberta per capita real GDP ranging from 0 percent to -.05 percent. By extracting the incremental rent from private owners of energy reserves, the provincial government can indulge in "province-building." By a combination of distortions of factor prices and direct purchase of public sector output, the government expands, relative to the classic staple boom, the size of the provincial economy. But government intervention, facilitated by collective ownership of oil and gas reserves, introduces a degree of "tragedy of the commons." Under any of the three options for government intervention, per capita provincial GDP is below that in the classic staple boom.

While the study by Norrie and Percy (1982) can be criticized for a number of dubious assumptions regarding the efficiency of private markets it does show that the issue of rent distribution is problematic. If rent is retained by the private sector as surplus profit it may leak from the regional economy or be dissipated in inefficiencies which can impede diversification. If it is used to fund economic development or social objectives in indirect ways such as obligations to undertake processing or maintain employment, it can be wasted building an inefficient industrial structure which collapses when the resource base is exhausted. If it is retained in general revenue it can be spent on dubious public programs which are difficult to curtail when and if resource rents decline. In this case, the public sector can get locked into an unsustainable level of expenditures.

While these issues are difficult to resolve, a few guidelines are in order. First, resource rents should be placed in a special fund such as the heritage funds in Alberta and Saskatchewan instead of accruing as general revenue. This will prevent highly volatile resource revenue from being viewed as a general revenue source suitable for financing ongoing governing programs. Instead, the funds can be used for special non-recurring expenditures or recurring expenditures based on some long term average revenue forecast. In the latter case, large surpluses would be accumulated during booms to cover deficits during downturns. Second, a proportion of resource revenue should be distributed directly to citizens as a resource dividend, and not remain at the discretion of politicians. The danger exists, and the public is aware of it, that the public bureaucracy may dissipate rent via self-aggrandizement, favours to special interest groups or provision of unwanted public services. If the benefits of efficient resource policy are transferred to the voter in a transparent manner, he or she will more directly experience the cost of blatantly sub-optimal policy and be more likely to oppose it.

One issue of rent distribution deserving special emphasis is the funding of an industrial strategy to diversify the economy. Many observers emphasize that diversifying the economy away from an excessive reliance on a limited resource base is essential to ensure long run economic health (Watkins, 1963, 1973; Richards and Pratt, 1979; Gunton, 1982). To do this the government must aggressively intervene to exploit potential linkages. Others maintain that attempts by government to expedite diversification beyond what the private sector is willing to undertake will simply dissipate rent funding uneconomic projects (Kierans, 1973). There is considerable evidence to support both positions. Clearly the private sector is not likely to identify or pursue all viable investments in a region, especially in a region with a weak entrepreneurial base dependent on external enterprise. To some extent, diversification can be expedited by reducing factor price distortions resulting from dissipation of rents in the resource sector. More direct public intervention may be required, however, to achieve economically justified diversification. By the same token, public intervention to force projects not otherwise pursued will inevitably result in some unjustified developments. Forcing development should only occur after a comprehensive analysis of the opportunities on a case-by-case basis.

10.4 CONCLUSION

The thrust of our analysis is that despite the problems associated with resource-based economies, the comparative advantage that Canada, particularly western Canada, currently enjoys in staples, and the potential gains from improved management, are so great that policy-makers should concentrate on rational resource management, and beware the seduction of more traditional industrial development. Central to our argument is that for both reasons of efficiency and equity, governments must give priority to collecting resource rents. As the figures in Table 10.2 illustrate, sound resource policy can have a dramatic effect on regional income.

We acknowledge efficient rent collection and rational resource policy to be a demanding task, fraught with decisions made under uncertainty and acrimonious political bargaining. It may appear we are inviting policy analysts to become Sisyphus, forever rolling their resource policy uphill, only to see it roll down at the end of the day. The consolation is that even incremental improvements will result in substantial gains in economic welfare.

11. INTERNATIONALIZATION AND THE SPATIAL RESTRUCTURING OF BLACK COAL PRODUCTION IN AUSTRALIA

Katherine Gibson

Over the past few decades the mode of black coal production in Australia has been radically transformed under the various impacts of international capital. Between 1970 and 1980, world output of black coal increased by 29 percent and exports increased by 52 percent as coal was heralded as the "bridge to the future" that would enable the developed world to ride out oil shortages until newer energy sources, such as nuclear and solar energy, could be expanded (WOCOL, 1980). The biggest increases in production came from non-traditional exporters such as South Africa and Australia (Rutledge and Wright, 1985, p. 307). In 1974 Rex Connor, then Federal Minister for Minerals and Energy, was to predict of the future "Australia will not be riding on the sheep's back. It will be riding in the coal truck" (Thomas, 1983, p. 11).

Certainly, Australia has now emerged as the world's largest coal exporter ahead of the US, Poland, South Africa, Canada and the USSR. Whereas in 1960-61 the output of salable black coal stood at 22 million tonnes of which just under two million tonnes were exported, in 1984-85 production was up to 118 million tonnes of which close to 84 million tonnes were sold overseas (Joint Coal Board, 1985, p. 7). For most coal producing countries, domestic consumption commands all but about 10 percent of total production. Australia, however, is exceptional in that about 65 percent of the country's black coal is exported. The massive expansion of production under the influence of international market and capital forces carved out a new geography of production very different from the economic landscape of coal mining which had previously existed. This chapter traces the evolution of a new spatial configuration of black coal production between 1960 and the mid-

1980s and explores some of the regional implications of this restructuring. In particular, the chapter considers only the coal fields of New South Wales and Queensland. Black coal production from other states is geared solely to domestic markets and at present constitutes a mere four percent of national output. Brown coal (lignite) production, which is concentrated in Victoria, is not considered as it is a wholly different commodity which possesses a separate market from that of black coal. This chapter also foreshadows the crisis that has occurred in the Australian local industry in the late 1980s (Gibson, 1990).

11.1 SPATIAL STRUCTURES OF PRODUCTION

The theory of "spatial structures of production" is useful in understanding the qualitative changes experienced by the Australian black coal industry over the last 25 years. While many geographers recognize that it is possible to distinguish a uniquely "capitalist" (as opposed to pre-capitalist) structure of industry and employment patterns (Buch-Hansen and Nielson, 1977), there is an increasing interest in differentiating the structures associated with different "phases of accumulation" (Webber, 1982) or "variants of capitalist production" (Gibson and Horvath, 1983). "Structure" is given to industry and employment patterns by those historically specific technical and social relations by which economic activity is organized. Change in spatial structures of production and spatial divisions of labour is seen to be brought about by the competitive dynamics of capital accumulation. These dynamics are themselves structured by phases of crisis during which qualitative transformation of the technical and social relations of production takes place (Gibson and Horvath, 1983). Structure and space are linked such that each restructuring introduces new spatial structures and modifies old ones:

> Each phase of accumulation establishes its own geography, which is a response to its logic of development in the context of an existing concrete distribution of forces and relations of production. That geography is manifest in direct production facilities, fixed capital, aids to consumption and the structure of class relations. Because of the need to socialise production, this capital is accumulated in only a few regions. Once that geography is laid down in the early years of a phase of accumulation, it is fixed by the long life and spatial immobility of the forces and relations of production (Webber, 1982).

FIGURE 11.1
COAL FIELDS OF EASTERN AUTRALIA
1954-55

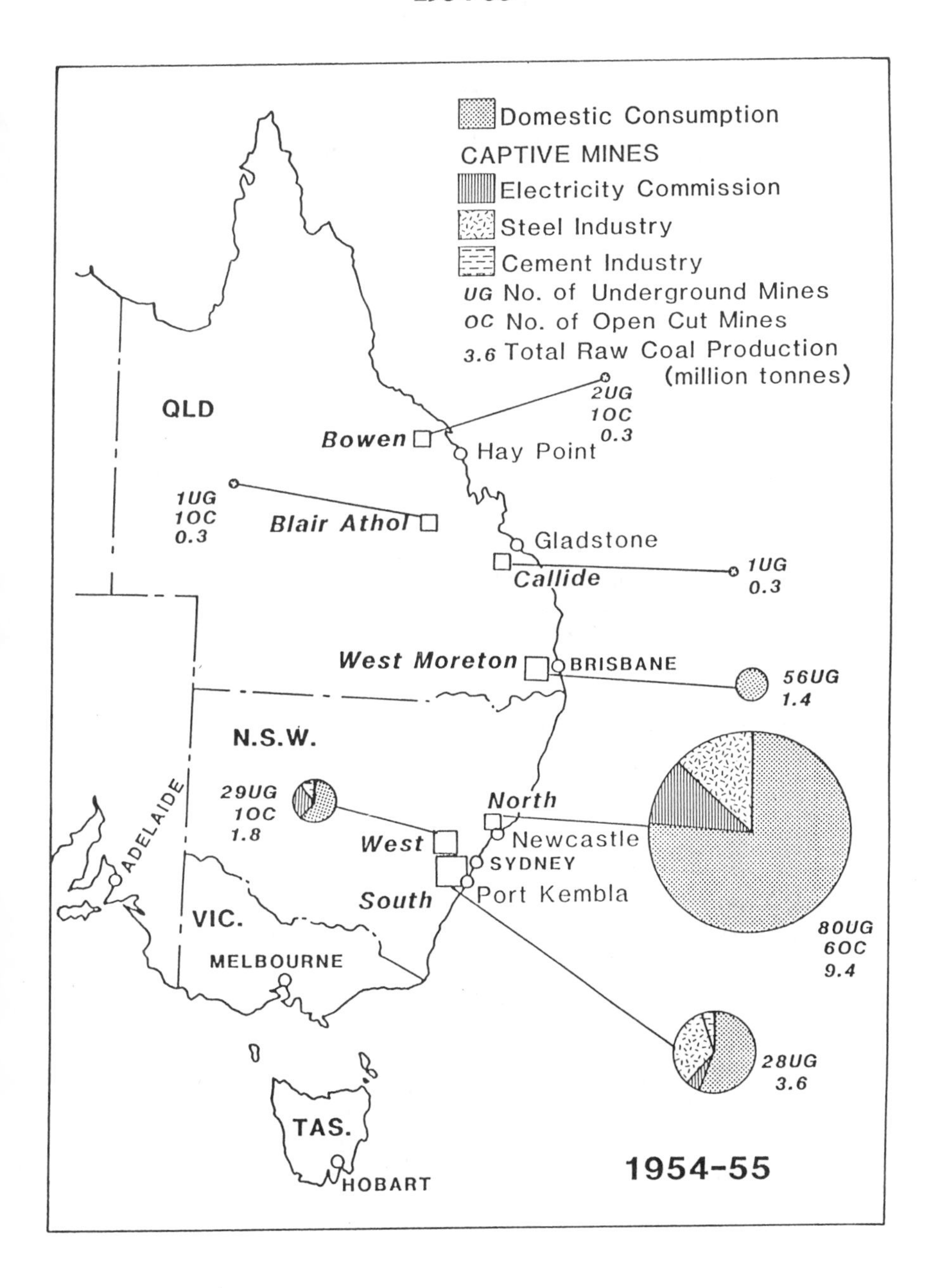

In this chapter it is argued that during the 1960s the existing spatial structure of coal production in Eastern Australia was transformed as a global variant of production was established within the industry (Gibson, 1984). The changed international linkages of the Australian black coal industry played a major role in creating a new competitive environment in which the widespread restructuring of production was inevitable.

11.2 THE SPATIAL STRUCTURE OF MONOPOLY VARIANT PRODUCTION

In the four decades prior to 1960, the market for Australian black coal was purely domestic. During this period the industry supplied the vital energy source for Australia's rapid industrialization, steaming coal for electricity generation, and for the declining needs of local shipping and railway transportation, as well as coking coal for the growing steel, metal refining and cement industries. Production was centred upon New South Wales (NSW). In 1950, of the 262 black coal mines of Eastern Australia, the 176 that were in NSW produced 80 percent of the national output (NSW and Queensland Department of Mines, 1950). Coal from NSW was exported around the coast to all the other states. Spatially, the industry was concentrated primarily in three coal fields outcropping at the edges of the Sydney Basin which had become well established during the nineteenth century (Figure 11.1).

During the interwar period a bifurcation in the organizational structure of the industry had taken place such that by 1950 the majority of mines were owned by small, independent, highly competitive local companies, financed internally by their share subscribers and operated at grossly inefficient levels of production. A minority of mines were owned as "captives" to large national coal consuming manufacturing companies, such as the monopoly steel producer, Broken Hill Proprietary Limited (BHP), and state institutions, for example, the NSW State Electricity Commission. This interest in the coal industry had been secured by state and private monopoly capital during the interwar period when the inefficiency and unreliability of production had forced the larger, more centralized consumers to buy their own mines. By 1954, 29 percent of NSW raw coal production was from captive mines owned by the State Electricity Commission of NSW, Kandos Cement Holdings Ltd. and BHP and two of its subsidiaries — Australian Iron and Steel and Blue Circle Southern Cement Limited, (Joint Coal Board, 1980, p. 12).

The majority of mines were underground and, owing to the thickness of seams, the predominant extraction technique employed was the Welsh "bord and pillar" method, one which leaves much coal unattainable underground (Mauldon, 1929; Gollan, 1963). In 1950 only 40 percent of coal cut in NSW was mechanically cut on average (compared to 70 percent in the United States at this time) and this average was 'pulled up' by the few very modern mines owned by the steel industry (Research Service, 1950, p. 62).

Prior to 1960 the spatial structure of coal production in NSW reflected the hegemony of the monopoly variant of capitalist production within the Australian economy at that time. As industrialization progressed within the protective policies of the federal government and with the direct assistance of state governments, the dominance of a monopoly variant of production was consolidated in selected manufacturing sectors and a segment of the coal industry. The spatial structure of a monopoly variant of capitalist production is characterized by the development of new, highly concentrated industrial agglomerations where all stages of the production of the commodity are located within a single geographical area (Massey, 1984). Industrial sites are often found at some distance from traditional industrial land uses where there is abundant space to set up large new factories structured around more efficient, but space extensive, techniques of production. Such a spatial structure necessarily relies upon significant state assistance in such forms as the development of new transport links, the rezoning of land for industrial use and the establishment of new residential areas and services.

The main coal fields initially developed by competitive capital in the late nineteenth and early twentieth centuries were all situated at some distance from the metropolitan centres of Sydney and Brisbane. In NSW industrial development by monopoly capital focussed upon the established coal fields. Under the impact of industrial investment, these "resource regions", previously structured around separated mining villages, blossomed into fullfledged industrial centres. The steelworks of Newcastle and Port Kembla attracted metal fabrication plants, engineering factories and munitions plants. Later, with the assistance of state tariff protection and decentralization policies, footloose industries such as clothing and textiles were established to employ the female population of work force age. The urban areas of Newcastle, Wollongong (respectively, the major cities north and south of Sydney) and Lithgow (on the western coal field) spread to encompass the old mining villages adopting a "dispersed city" structure (Daly, 1968). Evidence of the townscape associated with the competitive variant of coal production can still be found among the inter- and postwar sprawl of residential development in the suburbs of Newcastle and Wollongong (Jeans and Spearitt, 1980). In the classic multiplier model of orthodox economic geography, coal production, under the influence of monopoly capital, generated a set of diverse regional industrial complexes. By contrast in Queensland, where less than 15 percent of Australian black coal production was mined from three small fields which had expanded during the interwar period as a response to the frustration of reliance upon the disrupted NSW supply (Hince, 1982), the spatial structure of coal production was characterized by small mining communities centred upon a few locally owned mines.

In summary the spatial structure of black coal production prior to 1960 reflected a monopoly variant of production which, while economically dominant nationally, was spatially dominant only in NSW. Figure 11.1 illustrates this spatial structure by showing the number of mines in each region, their output and the domestic and corporate nature of consumption. The geography of monopoly capital was thus overlain and in fact limited to the initial investment pattern in coal mining associated with the earlier predominance of a competitive variant of production. In Queensland the monopoly variant was underdeveloped and the spatial structure of production reflected more the geography of isolated single company mining villages associated with the competitive variant.

11.3 THE SPATIAL STRUCTURE OF GLOBAL COAL PRODUCTION

After 1960 the geography of black coal production was to be radically restructured with the emergence of a global variant of capitalist production in the industry and a major change in international energy requirements. The process of internationalization had taken root well before the major shake-up of the global energy market precipitated by the oil price rises of 1973-4 and the subsequent reversal of the energy substitution trend of diesel for coal which had occurred some 40 years before. In the period 1860 to the 1920s, 30 percent of Australia's total coal output had been consigned to world markets, largely supplying British shipping interests in the Pacific region (Burley, 1960, p. 394). This international link ended with the decline of British shipping in the Pacific region during and after the First World War and the shift to diesel as the major transportation fuel in the interwar period.

The re-emergence of an international export market and the penetration of a largely domestically owned and financed industry by foreign transnational mining corporations took place as a result of two rather fortuitous developments. On the international front we saw in the late 1950s and 1960s the rapid industrialization of Japan and South Korea and their development of major domestic steel industries. Both economies were reliant upon imports of coal and iron ore to fuel industrial growth. On the domestic front the Australian coal industry, at this point mainly located in NSW, had just experienced a decade of significant reorganization and rationalization of the production and marketing of coal, instigated by the state through the authority of the NSW Joint Coal Board.

The reasons behind this unique experiment in quasi-nationalization of a large part of the non-monopoly sector of the industry from 1947 to the 1960s are documented elsewhere (Gibson, 1984; Gollan, 1963; Fisher, 1987). For the purpose of this chapter we need only recognize that international mining capital made its first entry into the Australian black coal industry via the acquisition of mines which had been taken over and modernized by the Joint Coal Board. The Canadian-based mineral multinational Placer Development was one of the first companies to profit from the rationalization instigated by the Joint Coal Board through its acquisition of a local contract mining company, Clutha Development. Figure 11.2 shows the company amalgamations which formed Clutha, as well as the takeovers in its ownership by international capital. Under Placer's management, Clutha absorbed into its corporate structure four state-owned or previously state-owned mines in the coal fields to the south and west of Sydney. A further state mine was added after D.K. Ludwig Universe Tankships, another multinational, took over ownership of Clutha from Placer. These state mines with their up-to-date technology and newly negotiated labour relations were tied into the rapidly developing Pacific export markets which the Joint Coal Board itself, through trade delegations, had secured.

Increasingly since the 1960s, ownership of the Australian black coal industry was internationalized through the acquisitions of foreign and Australian multinationals, both of which are involved in the global sourcing of coal to the power generation and steel industries of a range of West European and Asian nations. Financing of these interests has been supplied by borrowings on global money markets and equity shares distributed between Australian companies and financial institutions, and overseas interests. The changes in ownership within the Australian black coal industry which coin-

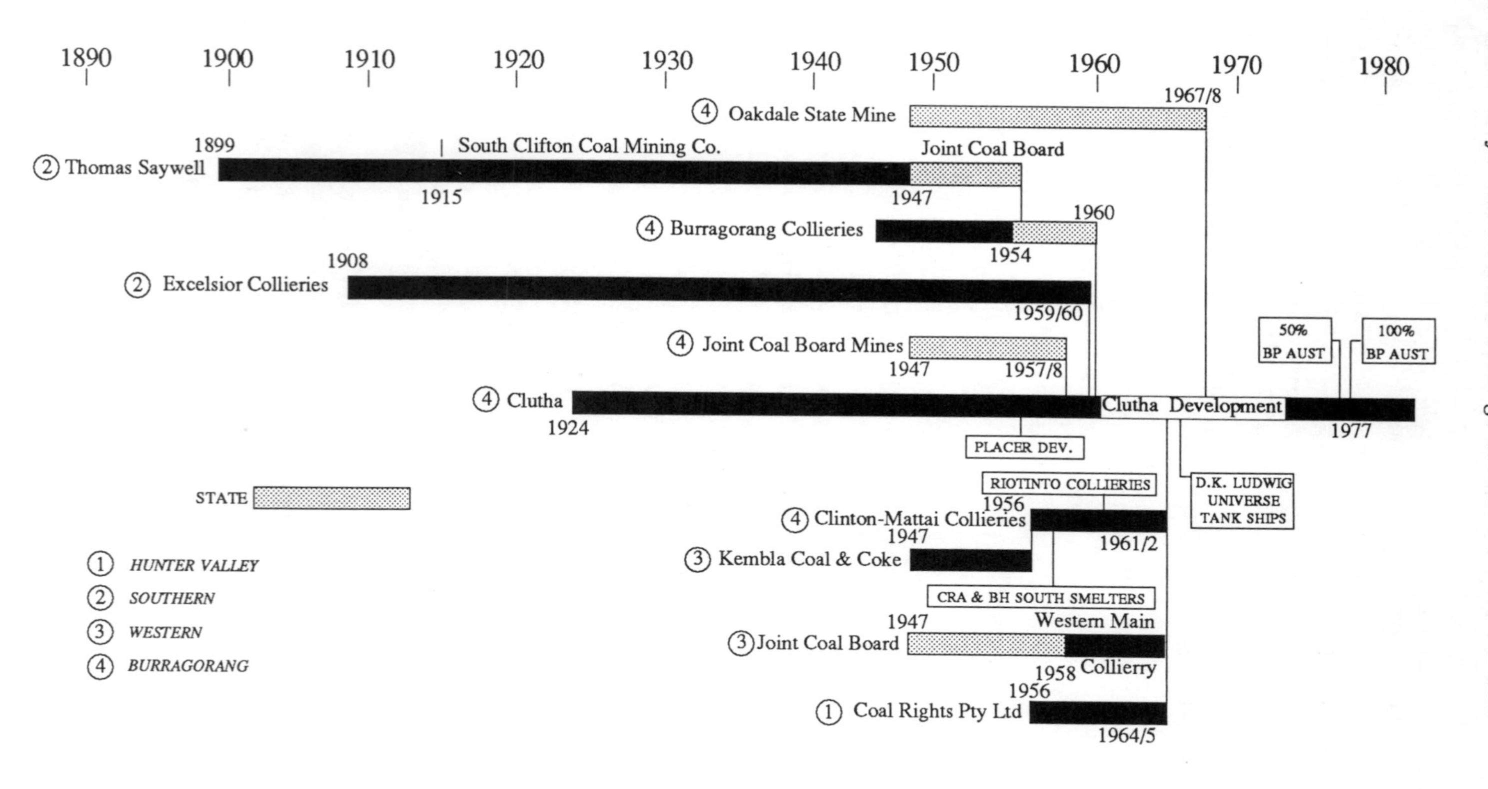

FIGURE 11.2
FORMATION OF CLUTHA DEVELOPMENT

cided with its increased international linkages represented a shift in command over production from independent locally based and owned coal companies to the highly centralized ownership structure of international mining, manufacturing and oil conglomerates, including both "foreign" and a growing number of "Australian" multinationals. The qualitative changes in the nature of coal mining that took place can be traced by looking at the new technical and social relations of coal production and the new spatial structure of the global variant which emerged.

Initially, international capital was attracted to buy up old mines producing coking coal which, before the period of state-led rationalization had been operating along competitive variant lines of production. These were all located in the old coal fields of NSW. Because of its higher quality coking coal, the south coast attracted a large share of this early round of pre-energy crisis international investment. In an attempt to continue the rationalization process begun by the state, the new multinationals encouraged the introduction of the most modern and expensive underground mining technology, the longwall mining unit. This technique allows a recovery rate of around 70 percent compared to an average of 52 percent from the continuous miners (Johnston and Rutnam, 1981, p. 16). However, the rate of change in underground technology has been relatively slow. In the 1970s, 87 percent of average daily output of NSW underground mines was still mined by continuous miners, 5 percent by longwall units and 8 percent was mined using coal cutters or by hand. By 1985 these first two proportions had changed to 76 and 24 percent respectively (Joint Coal Board, 1980, p. 63; 1985, p. 40).

Despite the upgrading of selected underground mines, by far the major technical transformation accompanying structural transition and changes in world demand was indicated by a shift in productive investment away from established underground mines to new open cut mines. In 1960-61, 90 percent (21 million tonnes) of Australia's total production of raw black coal was obtained from underground mines, but by 1984-85, while absolute tonnage increased (to 49 million tonnes), the proportion of coal from underground mines had dropped by almost two thirds to 33.5 percent (Joint Coal Board, 1985, p. 16). This dramatic change can be seen from Figure 11.3. A notable feature of the investment into Australian open cut development has been the prominance of oil companies. Since the oil shock of the early 1970s, the major oil companies have increased their role in the international coal sector, gaining a position of majority ownership in all the main export steaming coal projects which are coming on stream now and over the next ten years (Rutledge and Wright, 1985, p. 311).

The development of new open cut mines was limited to those areas further from the coast where seams are relatively shallow and gently dipping. Recent developments in the unit size and capacity of surface mining equipment, such as stripping shovels and walking draglines, have made much larger scale open cuts feasible; the depth at which open cut mining can be carried out has been increased as has the ratio of overburden to coal which can be profitably worked (Rutledge and Wright, 1985, p. 306). The type of open cuts developed during the 1970s allowed productivity, measured in terms of raw coal output per shift, to increase some threefold to fourfold over that of underground mining (Joint Coal Board, 1985, p. 70).

The depth of the coal seams on the south coast and the Burragorang Valley precluded open cut mining, but the other two older fields of NSW and the three established coal fields in Queensland all experienced new open cut investment in the 1970s and 1980s. However, the major spatial impact of open cut developments has been the opening up of new coal fields; Singleton-North West in NSW, and in Queensland the Mackay and Blackwater districts in the Bowen Basin, Callide and Kianga-Moura. Although the extent of coal reserves in these areas had been known for many years, most of these areas were, before the 1970s, primarily agricultural districts.

The new coal fields with their complex international linkages grew rapidly over the next decade. New labour forces had to be recruited, mining towns constructed, transport links built and mines developed. The character of these new resource areas could not have been more different from the close mining communities of NSW surrounded by a complex industrial fabric. To cater for the population influx, old one-horse towns such as Blackwater, Moura and Clermont were redeveloped and new company towns like Moranbah and Dysart were built from scratch. The skills required for open cut mining are relatively easy to learn and are possessed by many workers other than miners. Thus, the majority of workers joining the expanding industry were recruited from the construction sites of the area or from agricultural and provincial town occupations. A small proportion did migrate from coal mining areas in Queensland and NSW where employment opportunities were declining (Hince, 1982, p. 35). By and large the majority of workers in Queensland were young with no experience of mining unionism. In sharp contrast, the expanding mining work force of NSW usually came from families traditionally linked to mining and unionism. Despite the social and class differences in the make-up of the new mining work force compared to that of older regions, it was prevented from becoming a major divisive force within the national industry. One reason is to be found in the nature of unionism in Australia and in the coal mining industry in particular. The Miners' Federation is a national industry union which covers all face workers and unskilled underground workers (that is all but the electricians and engine drivers employed underground), as well as operators of heavy-duty coal hauling trucks and trades assistants in open cut mines. New mine workers in Queensland therefore joined the Miners' Federation and formed their own district branch called the Queensland Colliery Employees Union. Much to the surprise of companies, especially of the foreign multinational companies developing the new Queensland coal fields, the young and inexperienced work force quickly became as militant as their more class-conscious comrades down south. This was partly due to the heavy handed and inflexible industrial relations established by companies such as Utah which, before its takeover by BHP in 1982, was the epitomy of the corporate "ugly American" (Thomas, 1983, p. 246).

Thus despite the technical potential for a real subordination of labour to capital in the labour process — open cut mining, isolation in company towns, a new previously non-unionized work force — the social relations which developed were not overly differentiated from those characteristic of underground production. Indeed in terms of union militance in Queensland, the open cut mines have consistently experienced a higher percentage of shifts lost through industrial disputation over the period 1975 to 1985 than have the underground mines, with no discernible pattern of differentiation between older and newer coal fields (Queensland Coal Board, 1984-85). In this case the traditional cohesion and militance of the coal mining work force, related perhaps to the peculiar set of social traditions surrounding a dangerous occupation, has been able to challenge the tendency global capital usually has to peripheralize its work force. In another sense, however, namely the multiplier effects engendered by coal production, the spatial structure of production under the global variant of production is very different from that associated with a monopoly variant. With the exception of those coal fields established prior to the emergence of global capital in the industry, most new mining areas produce almost exclusively for export markets. Thus, apart from transport links to ports and port developments the local spinoffs associated with these developments are limited to providing services to the mining work force — hospitals, schools, retail and government functions. The international orientation of production has constituted these places as distinctly single-industry areas with very little hope for diversified development. Their economic vulnerability is, therefore, exacerbated.

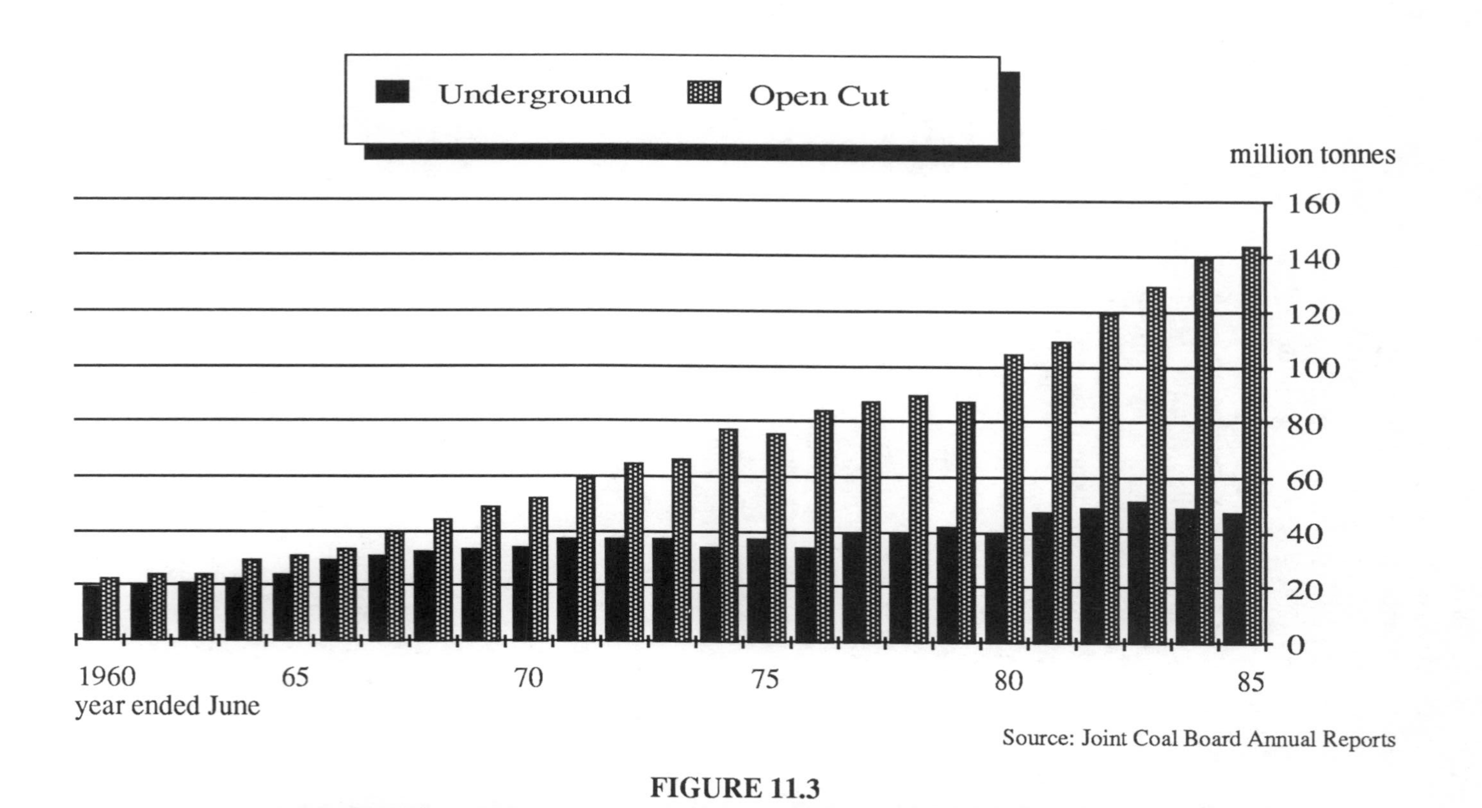

Underground
Open Cut
million tonnes
160
140
120
100
80
60
40
20
0
1960
65
70
75
80
85
year ended June
Source: Joint Coal Board Annual Reports

FIGURE 11.3

The spatial structure of Australian coal mining under a global variant of production is illustrated in Figure 11.4 which shows the type and number of mines in each district, raw coal production and its market destination — whether for export or domestic consumption. Although partially overlain on the investment structure established by monopoly capital and competitive capital before it, global capital has added wholly new spaces to the economic landscape associated with coal production.

11.4 CONCLUSION

Over the past few decades the international black coal mining industry has been somewhat shielded from the operations of the law of value because a unique temporary role as the world's primary industrial energy fuel was ascribed to it in 1973. For some time it enjoyed the investment status that the oil industry had possessed in the postwar period. Capital flowed unfettered into coal developments worldwide. In a classic attempt to cut production costs open cut developments were the big attraction, although the modernization of underground operations was also seen as a short-term strategy yielding immediate returns that allowed segments of the market to be cornered. The impact on the geography of coal production in Australia has been to create three different "types" of coal field or district, which comprises a physical entity defined by the spatial proximity of a grouping of mines. In that regions can be conceived as "types of places, [which are] not necessarily contiguous" (Webber, 1982, p. 2), each type of coal field constitutes a single region characterized by either growth, rationalization or decline. The region defined by growth includes the new and developing export-oriented resource areas structured by global capital such as the Singleton-North-West field in NSW and the Mackay, Blackwater and Callide fields in Queensland. The region defined by rationalization includes the older industrially diversified regions which are restructuring under the impact of the transition to a global variant of production such as the western and Newcastle fields in NSW and the Blair Athol and Bowen fields in Queensland. The region defined by decline includes the other older industrially diversified regions which are experiencing employment and investment decline, (under the impact of the same global transition), for example, the two southern fields in NSW and the Kianga-Moura and West Moreton fields in Queensland.

Despite warnings of an imminent crisis of oversupply in the early 1980s as oil prices dropped, companies have continued to pour investments into

FIGURE 11.4
COAL FIELDS OF EASTERN AUSTRALIA
1984-85

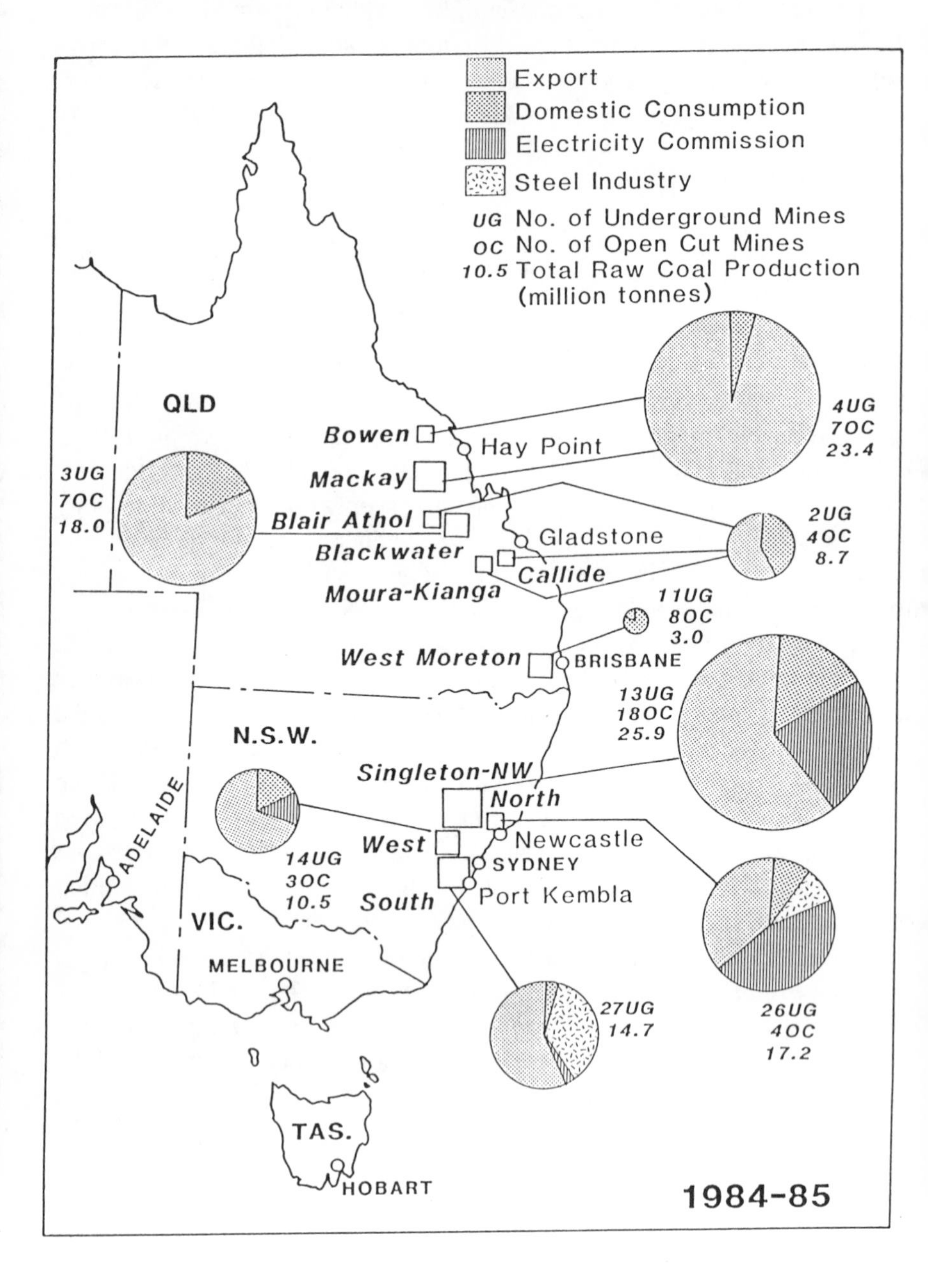

new developments. While many large-scale projects in the United States, Canada, South Africa and Australia are just beginning to come on stream, more development investment channeled through the same companies operating in these economies is flowing to the Third World. Two huge open cut mines are currently being developed by Exxon in Colombia on the El Cerrejon field where there is an estimated 2.5 billion tonnes of economically recoverable reserves (Rutledge and Wright, 1985, p. 313). BHP is currently developing a huge open cut mine in Indonesia on the island of Kalimantan which will come on stream soon (Byrnes, 1986). These developments can only exacerbate the already acute situation of international oversupply. Inevitably, the price of coal is dropping down to a level which is more reflective of its true value, for in real terms the average socially necessary labour time involved in the production of coal has been drastically cut over the last two decades. The underlying operations of an international law of value are becoming apparent, much to the dismay of producers in Australia. South Africa is threatening to sell its coal at $US25 a tonne, and has forced a drop in the free market spot quotes of Australian coal from $US33.50 to US$29.50 a tonne (A$42.15) (Behrmann, 1986, p. 19). Not surprisingly, the Australian government is being accused by the South Africans of economic self-interest in its support of sanctions against South African trade.

These international developments have many implications for the Australian coal fields. To some extent they herald the widespread introduction of an element of economic vulnerability, which selected older fields have already felt even during the boom period. For the operations of the law of value are not unfamiliar to those districts, such as the south coast, which were not able to take part in the slashing of production costs which accompanied open cut development in Australia. Clearly, the unexpected period of bounty experienced by all coal mining areas during the 1970s has ended and a more complex phase of rationalization and scaled down development has begun in which all three types of region will suffer.

ACKNOWLEDGMENTS

I would like to thank John Roberts for his excellent cartographic assistance in the preparation of this paper, and I would like to acknowledge the pilot research assistance provided by Sydney University second year students taking my Economic Geography course in 1986, in particular Joanne White and Kathleen Mee whose research findings have stimulated my thoughts.

12. INDUSTRIAL RESTRUCTURING IN THE AUSTRALIAN FOOD INDUSTRY: CORPORATE STRATEGY AND THE GLOBAL ECONOMY

R.H. Fagan and D.C. Rich

This paper analyses the Australian food industry to explore the dynamics of industrial restructuring in the decade or so since the mid-1970s. Both the industry's geography and its organizational structure are examined. The analysis is both extensive, in which sectoral and spatial changes are identified, and intensive, in which individual corporate strategies are examined in a search for underlying processes which can be generalized (Sayer and Morgan, 1985). Detailed study of corporate strategies for the largest firms in the food industry reveals the relationship between regional industrial change and internationalization of capital. The geographically uneven impacts of industrial change in this sector result from the interplay between global processes and the particular circumstances of Australia's industrial structure.

12.1 INTERNATIONALIZATION OF CAPITAL AND INDUSTRIAL RESTRUCTURING

During the 1970s transnational corporations (TNCs) began to employ a new international division of labour (NIDL), such as that described by Frobel et al. (1980), as a response to falling rates of profit in established markets and subsequently to economic crises (see Bradbury, 1985; Gibson and Horvath, 1983). This trend involved deindustrialization in parts of the United States and western Europe, and new forms of dependent industrialization in a group of Third World countries, along with the decomposition of production and deskilling of work force requirements to take advantage of low-cost labour pools (Cohen, 1981; Fagan et al., 1981). The "new" feature of NIDL was the growth of export-oriented manufacturing in the newly-industrializing countries (NICs) of peripheral Europe, Latin America, east and southeast Asia, and the search by capital for a worldwide labour pool for use in recreating conditions for accumulation (Trachte and Ross, 1985).

Despite the initial attractiveness of this NIDL thesis, more detailed empirical analysis of the underlying processes has exposed flaws in its conceptions of relations between deindustrialization in the core and the changing global economy (Browett, 1985). First, it places too much stress on the "offshore" exploitation of low-cost labour by TNCs. Movement of standardized production to low-wage locations has been concentrated in particular sectors such as clothing, motor vehicle production and electronics, and even in these industries, has not been the only strategy employed by capital. Second, it severely understates the role of both the state and local capitalists in determining the geography of the NICs. While TNCs remain a principal institutional vehicle for internationalization of capital, it is important not to overstate their role.

Third, the role of technological change has been much broader than suggested by the neo-Fordist approach of the NIDL theory. New technologies have also been employed to facilitate industrial restructuring inside the major markets, accompanied by widespread reorganization of relations between capital and labour involving work practices. Fourth, there has been a massive growth in the scale of the world financial system since the late 1970s (Andreff, 1984; Coakley, 1984; Daly, 1984; Daly and Logan, 1986). Large-scale loans have been employed not only by state and capital in the NICs but also to finance reorganization of manufacturing within and between the industrial core countries. This includes takeovers and interpenetration of major markets for which competition between the largest industrial organizations has become intense. This internal restructuring, including changes in technology and the labour process, has been more widely employed by capital to restore accumulation than the search for new sources of absolute surplus value in NICs (see Jenkins, 1984).

In addition to technological change, restructuring has involved various combinations of disinvestment and job-shedding at particular locations, and sometimes relocation to new production sites offering various cost advantages. Such reorganizations of production can also occur after financial restructuring as competition between large firms comes to reflect the struggle for control over capital rather than over specific product markets. Restructuring is also changing the relationships between large and small firms; for example, since the late 1970s there has been rapid growth of subcontracting (Holmes, 1986) and outworking (Mitter, 1986) in which some production moves "off-factory" rather than offshore (TNC Workers' Research, 1985).

Analysis at the urban and regional scale must be placed in the context of changing global industrial and financial systems. The argument presented here outlines a critical relationship between abstract theories of capitalist industrial restructuring, and empirical analysis of specific sectors, corporations and regions. Restructuring at the local scale "...includes contingent relations whose form cannot be theorised and hence known in advance, but which must be discovered empirically" (Sayer, 1985, p. 4). Yet adequate theory cannot be developed simply from a string of case studies. The two approaches must continuously enrich each other, with the structural relations and technologies of production setting the framework for human agency and contingent circumstances operating at specific times and places. Such a theoretical framework could explain the widely observed core-periphery patterns, which deepened in most industrialized countries during the long boom, *and* their replacement by more complex mosaics of uneven development since the mid-1970s.

12.2 CORPORATE RESTRUCTURING AND SPATIAL CHANGE IN THE AUSTRALIAN FOOD INDUSTRY

The Australian food industry provides a good case study of the organizational and spatial impacts of restructuring processes since the mid-1970s for five reasons. First, it is one of Australia's oldest manufacturing sectors and can be used as a mirror for major structural changes in Australian capitalism. Second, it is based on Australia's very large agricultural production capabilities. It presents an important study of key linkages between Australia's fluctuating fortunes as a major exporter of foodstuffs and the strategies of both capital and the state. Third, it is diverse in terms of commodities produced, technologies employed, levels of industrial concentration, relations between capital and labour, and the relative roles of large corporations (including TNCs) and small firms. Consequently, the pressures for, mechanisms of, and outcomes from restructuring are varied. Yet fourth, the industry is not substantially involved in tariff-protected, import-replacing production and pressures for restructuring have differed from those confronting the production of clothing and footwear, motor vehicles, and industrial and electrical equipment. Finally, it lies outside conventional product cycle explanations of the investment strategies of TNCs. Despite these features, it has experienced very rapid restructuring since the late 1970s.

Nationally, the food industry showed steady growth in output and employment during the 1960s and early 1970s. Between the early 1970s and the mid-1980s, the food, beverages and tobacco industry grouping was generally more stable and buoyant in terms of value added than the manufacturing sector as a whole. Employment in food, beverages and tobacco peaked at 204,100 persons in 1973-74, the same year as employment peaked in Australian manufacturing as a whole. During the mid-1970s, however, proportionately fewer jobs were shed from the food industry so that its relative contribution to manufacturing employment increased both nationally and in specific regions. By contrast, during the early 1980s, job-shedding reflected or was somewhat more extreme than rates for manufacturing as a whole. Compared to other industries the food industry was also distinctive in that job loss occurred steadily each year rather than in the major slumps experienced in basic metals, industrial and electrical equipment. A different pattern of restructuring is suggested.

12.3 CORPORATE STRATEGIES AND STRUCTURES: AN OVERVIEW

Towards the end of the long boom, the level of concentration and centralization in the food industry as a whole was amongst the lowest in Australian manufacturing. The largest 20 enterprises contributed only 33 percent of total turnover in the industry in 1971-72, although concentration was very much higher in some branches, such as sugar milling and refining, brewing and cereal production. Further, concentration and centralization trends were proceeding primarily in response to growing penetration by TNCs from the United States and western Europe including Unilever, British American Tobacco and George Westons Ltd. (see Burns et al, 1983; Fagan, 1987).

Similarly, in 1972 the overall level of foreign ownership and control appeared lower than for manufacturing as a whole. Yet the TNCs were more important than industry-wide figures indicated, as overseas firms commonly established dominance in particular product markets such as: breakfast cereals (Kelloggs); soft drinks (Coca Cola); and cheese (Kraft). In other cases, notably cakes, bread, confectionery and meat processing, they were market leaders in sectors which contained a large number of small enterprises. Most of the world's largest food TNCs established Australian subsidiaries during the 1960s and early 1970s sometimes entering the market through major acquisitions. Of the 30 largest TNC food manufacturers operating in Australia by 1980, 13 were based in Britain and 13 in the United States. In the same year, 48 of the world's largest food companies were domiciled in the United States and 22 in Britain (Burns et al., 1983).

Since the mid-1970s, there has been dramatic corporate restructuring in the Australian food industry involving both domestically owned and foreign firms. Merger and takeover became particularly frantic during the 1980s. Three features distinguish this wave from that which occurred in the 1960s. First, an increasing proportion of takeovers led to corporate diversification rather than simply increased market power. Second, and partly as a result, it has been more common for recent takeovers to be followed by extensive rationalization and selective divestment of either plants or entire divisions, as acquired companies have been integrated into the corporate profit-making structure. Third, the scale of recent takeovers has been fed by the high rates of foreign capital inflow, especially as large-scale loans. These acquisition waves broadly coincide with those discovered in other industrialized countries, and the latest period is associated with the restructuring crisis (Aglietta, 1979).

Restructuring during the 1980s took different paths in individual branches of the food industry. Agribusiness was reorganized rapidly with Elders, Dalgety and Industrial Equity emerging as the most powerful corporate controllers of agricultural sales (including exports). By the mid-1980s Elders IXL Ltd. controlled 40 percent of total livestock sales in a fiercely competitive industry containing thousands of small agents (Sargent, 1985, p. 75). A further 20 percent was controlled by the British TNC Dalgety. These two corporations also controlled some 84 percent of Australia's wool sales between them and became two of the largest participants in rural land and property sales. Conflict has arisen between these agribusiness corporations and marketing and regulatory bodies established by the state to oversee Australia's exports of primary products. Elders IXL is seeking a greater role in the marketing of Australian wheat, for example, currently managed by the Australian Wheat Board. By 1983, Elders IXL Ltd. controlled some 15 percent of Australia's rural export trade (Sargent, 1985, p. 68).

Meat processing also became more highly concentrated, with Elders IXL, Adelaide Steamship Company Ltd (beef and pork) and Amatil Ltd (chicken) forming key relationships with smaller producers. The rapidly growing domestic market for chicken meat became dominated by just two firms. By 1985 the independent Inghams Enterprises Pty Ltd. and Amatil (a British Tobacco subsidiary) supplied nearly 80 percent of the domestic market and monopolized the supply of day-old chicks to farmers (Australia: Prices Surveillance Authority, 1986). Similarly, brewing became dominated by two firms, Elders IXL and Bond Corporation Ltd., which over the past two decades progressively absorbed smaller regionally oriented firms and now supply the national market. Still other branches of the food industry, notably sugar, cereals and baking, became characterized by intense merger and takeover activity and increased levels of corporate concentration (see O'Neill 1986; Rich, 1987).

By contrast, deconcentration took place in parts of general food production (including fruit canning) and there was rapid disinvestment in some of the least profitable sectors, with some large groups selling off plants producing low-profit commodities to firms remaining in the industry. Canning, as a low-profit sector, was abandoned by corporate producers after export markets for canned fruits began to collapse in the 1970s. Henry Jones IXL (now part of Elders) sold its Kyabram (Victoria) canning plant and equipment to a farmer-workers' co-operative in the Riverland irrigation district of South Australia in 1979. The South Australian state government assisted with the purchase, while the Victorian government paid for the relocation of the plant to the Riverland. The canning industry is now left largely to four co-operatives and is under severe stress (Sargent, 1985). Divestment has increased centralization in margarine and frozen foods industries, but decreased it in some general food lines.

The 1986 trans-Tasman merger of Goodmans (NZ), Fielders Gillespie Ltd. and Allied Mills Ltd. is a good example of the complexity of recent financial reorganization as defence against future takeover and to secure dominant long-term positions in food markets. During the 1980s, Goodmans, one of New Zealand's largest agribusiness groups, engaged in a battle with Industrial Equity Ltd. (an Australian-based corporate raider with large New Zealand interests) and, in 1984, Elders IXL Ltd. defended Goodman against a takeover bid by purchasing a 22 percent shareholding. In return, Goodmans purchased Elders' Australian margarine and edible oil interests and made a significant investment in Elders IXL shares. In divesting food sectors lying outside its overall corporate strategy, Elders had attempted to sell this division to Unilever but was prevented from so doing by Australia's Trade Practices Commission.

In 1985, Allied Mills Ltd., one of Australia's largest margarine, flour milling, baking and general food groups, also became a target for corporate raiders. In the ensuing battle one of Allied's principal competitors, Fielders Gillespie Ltd., merged with Goodman and the new group mounted a successful takeover of Allied early in 1986. The new Goodman Fielders Ltd. had 53 percent New Zealand assets and the new board of directors became interlocked across the Tasman with Allied, Fielders and Elders appointing directors. Apart from further centralizing baking and flour production, the merging of Goodman and Allied Mills gave Goodman Fielders a majority share of the Australian margarine market. Again under pressure from the Trade Practices Commission, Goodman sold the former Elders' margarine division to the New Zealand Dairy Corporation late in 1986. This division and its Queensland production plants have been owned successively by Provincial Traders Ltd., Henry Jones, Elders IXL, Goodman Ltd. and finally the Zealand Dairy Corporation, illustrating the complex restructuring arising from interactions between corporate strategy and state regulation.

The more complex corporate structures developing during the 1980s bore a direct relationship to both increasing technological sophistication in food manufacturing and rapidly increasing promotional expenditures in national markets. These trends were comparable with those in the United Kingdom and the United States (Burns et al., 1983). Another important force for reorganization in food manufacturing has been restructuring among the large food retailing groups. Direct buying by the large supermarket chains and the diffusion of generic brand-names became more common in the 1980s, and recent mergers have sharply increased centralization in food retailing.

In summary, since the mid-1970s four broad corporate strategies for restructuring the conditions for profitability and capital accumulation can be generalized. First, there has been continued centralization of control in food commodities where competition for share of local markets has become more intense. Acquisitions to increase vertical integration have been important in most of these sectors. Some companies have sold out of certain commodities to concentrate on other activities.

Second, some well-established agribusiness corporations have diversified away from their food production bases to cope with long-term price falls and instability in international markets. Examples include: the sugar millers and refiners, which moved into minerals and energy, and the production of chemicals and building materials; and Elders IXL Ltd., which diversified into manufacturing, financial and business services.

Third, some corporations have diversified selectively into food production to provide a stable cash flow as a basis for long-term accumulation strategies in other sectors. Examples include the Bond Corporation Ltd., which emerged after 1985 as the only national competitor for Elders IXL Ltd. in the Australian brewing industry.

Fourth, a recent strategy has involved Australian companies using their base of agribusiness, food and beverage production to launch strategies for the internationalization of capital transferring new investments into major northern hemisphere markets. Examples include: Elders IXL Ltd. in the United Kingdom and Canada; Bond Corporation in the United States and the United Kingdom; Industrial Equity Ltd. in the United States; Goodman Fielders Ltd. in the United Kingdom; and proposals by Arnotts Ltd. to commence production in the United States. Some of the target companies have been among the largest in the host-country, either nationally or regionally.

The corporate reorganization has been made more complex by the corporate raiders such as Bond Corporation Ltd., Adelaide Steamships Ltd. and Industrial Equity Ltd., which built up financial structures which included some key positions in the food sector. Takeover bids for undervalued companies have been based partly, or largely, on loan funds and divisions of the company were commonly sold off after acquisition to reduce the total debt. During the 1980s all of the restructuring strategies listed above involved increased use of finance capital raised on the Euromarket and in North America.

Centralization of capital in the food industry has contributed to further increases in the control over Australian industrial production exercised from Sydney and Melbourne (Taylor and Thrift, 1980; Fagan, 1987). Yet headquarters of the largest food companies are relatively more dispersed than for Australia's largest business organizations generally. This results partly from the survival of large food companies in the peripheral states and partly from the activities of the corporate raiders based in Adelaide and Perth. Their activities have caused a net geographical dispersal of ultimate control in some important branches of production since 1980. In terms of production and employment, there have been quite uneven effects on different sectors of food production and in different locations. In many sectors, the corporate reorganizations have been too recent for locational trends to have become apparent.

12.4 EMPLOYMENT CHANGE

Capital's relations with labour and the state are critical in the restructuring strategies. Job-shedding in the food sector has involved intensification, or changes in labour practices; widespread investment in labour-replacing technology; and both sectoral and spatial rationalization. Technological change has been very important among the largest firms although it has often followed post-merger rationalization and disinvestment, for example in brewing, baking and general foods.

During the 1980s, disputes arose within branches of the food industry over attempts by capital to introduce new labour practices. The most celebrated, which became a test case for the introduction of new labour contracts, involved the small Mudginberri abattoir in the Northern Territory and resulted in costly defeat for the union concerned. The company was assisted in prosecuting the case by the newly formed National Farmers' Federation, the president of which at the time was also a director of Elders IXL Ltd., which owned a large meatworks in the Territory.

The geographical pattern of production and employment in the food industry remained throughout the 1980s essentially one of concentration, with the bulk of the industry located in the five mainland state capitals, notably New South Wales (NSW) and Victoria. Despite a superficial resemblance, there are some important differences from the geography of manufacturing as a whole which can be understood only in the light of the restructuring processes outlined above. In 1983-84, for example, 59 percent of employment in the food industry was located in the five mainland state capital cities compared with nearly 75 percent of employment in manufacturing as a whole. In every mainland state, the food industry is less geographically concentrated in the metropolitan area than is manufacturing generally. The difference is greatest in South Australia (where wine and fruit processing remain important) and Queensland (where meat and sugar predominate).

TABLE 12.1

THE GEOGRAPHICAL DISTRIBUTION OF THE AUSTRALIAN FOOD INDUSTRY, 1984-85, COMPARED WITH MANUFACTURING AND POPULATION[1]

State/Territory	Employment[2]			Value added[3]			June 1985
	Food (no.)	Food (% of total)	Manuf. (% of total)	Food ($ million)	Food (% of total)	Food (% of total)	Population (% of total)
New South Wales	51,7671	31.0	35.8	2,175.1	31.8	36.6	34.8
Victoria	49,651	29.7	35.0	2,203.4	32.2	34.3	26.2
Queensland	31,567	18.9	10.8	1,227.0	18.0	11.4	16.2
South Australia	15,284	9.2	9.1	528.9	7.8	8.1	8.6
Western Australia	11,990	7.2	6.3	453.7	6.6	6.5	8.9
Tasmania	5,752	3.4	2.4	212.2	3.1	2.5	2.8
ACT	489	0.3	0.3	15.6	0.2	0.3	1.6
Northern Territory	454	0.3	0.3	19.3	0.3	0.3	0.9
Australia	166,954	100.0	100.0	6,835.2	100	100.0	100.0

Source: Derived from *Australian Bureau of Statistics, Manufacturing Establishments: Details of Operations, 1984-85*, various states.

Notes:

At a state level, NSW and Victoria accounted for 71 percent of Australian manufacturing employment in 1983-84 but only 61 percent in the food industry (Table 12.1). Food processing loomed largest in the industrial mixes of what are generally regarded as peripheral states. Yet clear core-periphery patterns break down when a different point of comparison is used. On a per capita basis, the food industry is least important in NSW, Western Australia and the two territories, and most significant in Victoria, Queensland, South Australia and Tasmania. Each of these groups includes parts of what are commonly thought of as Australia's core and periphery, indicating the importance of identifying core-periphery structures empirically and grounding interpretation in the underlying structures of production.

A tendency is evident at several levels for the food industry to have become more spatially dispersed between 1971-72 and 1983-84 (Table 12.2). First, at the state level, the proportion of food industry employment located in NSW and Victoria declined by 3.5 percentage points as the number of jobs in the two states fell by 26,479 (or 20 percent), compared with 5,625 (nearly 8 percent) in the rest of Australia. Similarly, while food industry employment declined by 22 percent in the five metropolitan areas, the decline was under 6 percent in the rest of the country. There were, however, sharp interstate differences in the extent of this trend, with proportional losses greatest in Melbourne and Adelaide and least in Brisbane among the state capitals. Third, at the intrametropolitan scale, the food industry has experienced the worldwide trend for inner city concentrations to be dissipated and be replaced by suburban clusters of manufacturing. Sydney illustrates this phenomenon. Between 1971-72 and 1983-84, food industry employment in the three inner city local government areas of City of Sydney (including South Sydney), Marrickville and Leichhardt declined from 16,696 to 7,404, that is, from 37 percent to 21 percent of the metropolitan total. There is evidence, within metropolitan Sydney at least, that while the food industry was relatively more clustered than manufacturing generally in inner city areas in the late 1960s, the trend to dispersal was more marked than in other industries during the 1970s and 1980s (Rich, 1987).

Such sizeable spatial shifts occurring in the context of centralization of control and job-shedding with little change in aggregate production can only be understood in the light of the strategies for corporate restructuring outlined earlier. The sharp fall in employment in NSW and Victoria, and in most of the state capitals, reflects corporate responses to market conditions via capacity cuts, moves into new products, amalgamation of plants and

TABLE 12.2

GEOGRAPHICAL VARIATIONS IN FOOD INDUSTRY EMPLOYMENT CHANGE, 1971-2 TO 1983-84[1]

Area	1971-72	1983-84	Change	% Change
New South Wales	66,412	52,855	-13,557	-20.4
Sydney	45,420	35,701	-9,719	-21.4
Rest of New South Wales	20,992	17,154	-3,838	-18.3
Victoria	62,805	49,883	-12,922	-20.6
Melbourne	44,661	32,920	-11,741	-26.3
Rest of Victoria	18,144	16,963	-1,181	-6.5
Queensland	33,932	32,115	-1,817	-5.4
Brisbane	15,675	14,537	-1,1138	-7.3
Rest of Queensland	18,257	17,578	-679	-3.7
South Australia	17,810	15,885	-1,925	-10.8
Adelaide	11,943	8,946	-2,997	-25.1
Rest of South Australia	5,867	6,939	+1,072	+18.3
Western Australia	13,600	11,776	-1,824	-13.4
Perth	10,903	8,442	-2,461	-22.6
Rest of Western Australia	2,697	3,334	+637	+23.6
Tasmania	5,981	5,925	-56	-.09
Australian Capital Territory	601	475	-126	-21.0
Northern Territory	449	572	+123	+27.4
Five metropolitan areas	128,602	100,546	-28,056	-21.8
Rest of Australia	72,988	68,940	-4,048	-5.5
Australia	201,590	169,486	-32,104	-15.9

Source: Derived from Australian Bureau of Statistics, *Small Area Statistics by Industry, 1971-72* and *1983-84*, various states.

Note:

1 1983-84 figures exclude single-plant enterprises with fewer than four employees, reducing apparent employment by 1.1 per cent.

adoption of new technology. Likewise, at the intrametropolitan level, the widely discussed "urban" processes of dispersal such as land supply, changing space needs, improved transport, and decentralizing markets, suppliers or labour (see, for example, Cardew and Rich, 1982; Rich, 1987) have been reinforced by corporate restructuring. Older inner city plants have been most vulnerable to rationalization. Indeed, even though the impacts of financial reorganization have not fully worked through to the level of production, such restructuring processes are beginning to outrun the longstanding mechanisms of intrametropolitan dispersal. In the two years to 1983-84, employment decline in the food industry was substantially more rapid in the Western Sydney Statistical Subdivision (which contains most of the city's industrialized middle and outer suburban zones) than in the metropolitan area as a whole.

At the interstate, intrastate and intrametropolitan levels, ownership centralization, amalgamation of plants and rationalization have tended to produce net geographical dispersal of food processing. Within non-metropolitan areas, on the other hand, there has been a trend to greater spatial centralization as widely dispersed, resource-oriented plants are closed, and production either reduced or centralized in fewer, larger premises. The sectoral unevenness of change is repeated, however, at this level with the substantial shake-out of dairy factories contrasting with the relative stability of sugar milling.

12.5 ELDERS IXL LTD. DOMESTIC RESTRUCTURING AND GLOBAL EXPANSION

Founded in 1839, Elders Ltd. grew through a series of mergers as an Adelaide-based pastoral house controlling a major share of Australia's rural exports by the 1970s. While these traditional activities dominate the company's annual sales revenue, Elders embarked on a program of restructuring as the profitability of Australia's primary export activities slumped. Figure 12.1 shows the results of this dramatic reorganization and demonstrates clearly how restructuring within capital has blurred the distinction between sectors.

In 1981, Elders Ltd. merged with the food processing conglomerate Henry Jones IXL Ltd. The asset backing of the historic pastoral house had become an attractive takeover target and the merger was primarily a defence against a takeover bid by the Perth-based Bell Group. Locked firmly into the

Melbourne corporate and financial establishment through Henry Jones, Elders gained new corporate management and the means to become an integrated agribusiness enterprise. Henry Jones IXL Ltd. gained a rich base on which to launch an ambitious expansion strategy. As a principal matchmaker in the defence, Australia's largest brewing group (CUB Ltd.) took a 49 percent shareholding in the new Elders IXL as a major diversification but was itself acquired by Elders in 1984 to complete its transformation into an agribusiness and manufacturing conglomerate. This had earlier included acquisition of the British Wood Hall group, a diversified pastoral, construction and trading company with interests in the Asia-Pacific region. Absorption of Wood Hall further centralized control over Australia's rural exports. In 1983, Elders took over F.J. Walker Pty Ltd. which gave it a chain of meatworks in eastern Australia and control over Mayfair ham and bacon products and made the company the largest single pig producer by 1984.

Elders' strategy had become clear by 1984-85. First, rationalization of the new corporate empire continued in the late 1980s and included the sale of lower-profit food processing divisions to other corporations. In addition to its withdrawal from margarine production, Elders IXL sold its general foods activities to Petersville (an Adelaide Steamship Company Ltd. associate) which, to satisfy the Trade Practices Commission, subsequently sold the former Elders frozen potato division to Canadian TNC McCains Ltd. Second, the rich assets and sales revenues from the breweries and hotel chains have become very important to the highly geared Elders IXL, as they have to Bond Corporation Ltd. Increasingly, the profits earned by the brewing division (46 per cent of Elders' total in 1985-86) are the basis for the corporate diversification strategy.

As the leading pastoral company, Elders has long-standing rural banking, insurance and real estate interests. These activities have been extended further into advanced services and the information economy. In 1984, the company introduced a computer-based information service for farmers. In addition, Elders IXL Ltd. increased its international activities not only in relation to Australia's trade but also as a global wholesaler, arranging trade deals between other countries. This has been accompanied by consulting activities and merchant banking in both Australia and the Asia-Pacific region. In 1984, Elders acquired the Private Investment Company of Asia Ltd., a merchant bank with strong involvement in financing Asia-Pacific trade. By 1984, Elders IXL Ltd. had applied for a full retail bank licence as part of the restructuring of the domestic financial sector. While the pastoral division

FIGURE 12.1
ELDERS IXL LTD.

ELDERS IXL LTD (HQ: Melbourne)

ECONOMIC SECTOR / CORPORATE DIVISION	PASTORAL	BREWING	'INTERNATIONAL' (Agribusiness and manufacturing)	FINANCE	OTHER INVESTMENTS (inc. Elders Resources Ltd 48%)
AGRICULTURE	Wool sales Livestock sales Breeding services		Wool trade Grain trade Meat export Hides and skins Intensive piggeries Stock feed Animal breeding Pineapples (Sth Africa) Hops		
MINING					Oil & gas exploration and gold (20%) Coal mining (50%)
MANUFACTURING		Brewing inc. Courage Ltd (UK)	Grain milling (UK) Wool processing Meatworks Ham/bacon Smallgoods Maltsters		Goodmans (NZ)(14%) South Australian Brewing Co. Ltd (20%) Automotive safety equipment; accessories B.H.P. Ltd (20%)
BUILDING & CONSTRUCTION					Hornibrook Elders CED
WHOLESALE & RETAIL	Rural supplies	Wines & spirits merchants	Global wholesaling Timber sales		Yates (NZ) (20%) rural services
FINANCE and PRODUCER SERVICES	Insurance Real estate Information services	Hotels	Shipping and chartering Engineering (Asia)	Merchant Bank Retail banking Real estate financing Stockbroking Travel Agents	'Jam Factory' (35%) (shopping centre complex)

contributed most to the company's sales revenue (45 percent in 1985-86), it contributed only 9 percent of corporate profit by 1985-86. The finance division was responsible for only 7 percent of total sales but 20 percent of total profit (Elders IXL Ltd., 1986).

Clearly it has become more profitable to provide finance, information, and other inputs to Australian farmers than to sell their produce on world markets, although farm output is central to corporate strategy so that farmers will require finance and information. The internationalization of Elders IXL Ltd. has involved the further centralization of control over Australia's trade, heavy dependence on international finance capital, and future corporate strategy aiming at offshore production especially in brewing.

In 1986 Elders IXL paid A$3,320 million to acquire Courage Ltd., Britain's sixth largest brewer with three breweries, over 5,000 tenanted or managed hotels, and a chain of liquor wholesale and retail outlets. In 1987 Elders further internationalized its brewing division by taking over Carling O'Keefe Inc., Canada's largest brewing group. Earlier in 1986 Elders had become entangled in a battle for control of Australia's largest industrial organization, BHP Ltd., and by April 1986 had emerged with a 19 percent shareholding of this very much larger company, purchased on-market largely with borrowed funds. Elders IXL and BHP between them control about one quarter of Australia's total exports. Almost immediately, BHP Ltd. outlayed A$1,200 million to purchase 20 percent of Elders IXL Ltd. Ironically, BHP argued that this investment would help Elders to finance global expansion (BHP Ltd., 1986, p. 11), in other words, to provide a more stable base of equity funds for takeovers in the United Kingdom. In January 1988, as part of a massive "buy-back" of its own shares, BHP Ltd. purchased control over the 19 percent shareholding held by Elder IXL Ltd. To help raise finance for this move, BHP announced it would sell back its investment in Elders to an associate of the agribusiness and brewing giant.

The example of Elders IXL Ltd. provides a mirror for: recent structural change in Australia; the crucial links between production, finance and advanced services; and the integration of Australia into the world economy through corporate networks. The Australian food and beverage industry has become the basis for global expansion of Elders IXL Ltd., and since 1980 its corporate strategies have played a major role in the restructuring of both production and financing of Australia's agriculturally based industries. Elders has invested in the NICs of the Asia-Pacific region but largely in financial and other services for the expanded industrial and trading opportunities. Direct investments in production have been directed largely to the United Kingdom, Canada and New Zealand, and have been accompanied by considerable rationalization in domestic brewing and vigorous competition with the other large corporation remaining in this branch of the food industry.

12.6 CONCLUSION

The Australian food industry demonstrates the importance of global processes in the industrial restructuring of the past decade. The nature of these processes varies for specific branches of production and their impact in particular cities and regions is determined by their interaction with local relations of production. The Australian food industry began restructuring later than other key manufacturing sectors but change was rapid throughout the 1980s. Although the organizational changes within capital are very recent, it is already possible to trace the effects of their interaction with labour and the state through to local change.

The search for a more adequate theory of industrial restructuring must bridge the gulf between abstract "top-down" theories of change in capitalism and concrete expressions of restructuring at regional and local scales. The internationalization of capital, rather than simpler notions of NIDL, provides the framework for understanding the roles of capital, labour and the state in regional and local change. The Australian food industry developed within the overall context of food trade with other industrialized countries and, after 1960, the impact of direct investment by transnational food corporations. Yet since the mid-1970s, both the TNCs and Australian corporations have restructured production locally, while the largest Australian firms have become more thoroughly integrated with the world capitalist system. Finance capital now plays the crucial role in spreading the restructuring through different branches of production as well as into particular cities and industrial regions.

ADDENDUM

Since this chapter was written, the most important trends identified in the empirical analysis have intensified. Financial restructuring of the major food corporations has taken an even more volatile path, especially following the stock market crash of late 1987. In particular, the difficulties of servicing very large debts in a period of high interest rates have come to dominate the strategies of many businesses.

Increasing centralization has continued to involve the largest food manufacturers in both Australia and New Zealand, Goodman Fielders Ltd., for example, merged with Wattie Industries and consolidated its control over frozen food and cereal production in New Zealand. CSR Ltd. increased its dominance of the Australian sugar industry by taking over the second largest milling and refining company, Pioneer Sugar Ltd. Adelaid Steamships Ltd. strengthened its control over Australian frozen foods, vegetables and fish products while, early in 1990, mergers dramatically increased the market share of the two largest wine-makers. Yet counter trends also continued as corporations continued to sell assets either to rationalize complex production and marketing structures or simply to raise cash to service their growing debt to transnational banks. Thus, Elders IXL Ltd. sold its leading position in pig meat production to its major competitors, thus divesting assets it had held for just three years. Small enterprises also were able to develop market niches in competition with the corporations. In brewing, sluggish market growth continued to intensify the struggle between corporate giants for control over the steady cash flows but their overall market share was challenged by new regional brewers and the "boutique" brewers developing specialized market niches.

The internationalization strategies pursued by the largest corporations earlier in the 1980s were also continued later in the decade. The food corporations sought to acquire key positions in North American and European markets, often attracted by the promise of large cash flows. For example, Elders IXL Ltd. merged its Canadian brewing operations with those of the Molson Corporation to create Canada's largest single brewing group. Likewise, control over commodity trade became even more important to the agribusiness concerns such as Elders which expanded its grain-trading activities from a base in the United States. In addition, all of the largest food corporations borrowed heavily from transnational banks both to fund domestic restructuring and their offshore activities. Yet during the late 1980s there were also some spectacular failures. After a bitter struggle, Goodman Fielder Wattie Ltd. withdrew a leveraged bid for the British food giant Rank Hovis McDougall PLC after surviving a counter-bid from that corporation. Elders' takeover bid for Scottish and Newcastle Breweries Ltd., designed to further consolidate its position in the British brewing industry, was blocked by the Monopolies and Mergers Commission because of the threat posed to competition. Bond Corporation Ltd., sustaining major losses in its United States brewing operation, finally withdrew from an almost disastrous pursuit of the British TNC Lonrho.

Such failures were symptomatic of a more turbulent competitive and financial environment and much of the frantic activity among the food corporations was impelled by financial struggles. Debt financing of both domestic and global strategies increased as corporations tried to defend themselves against leveraged takeover bids financed by transnational banks. By the end of the decade, however, many of these attempts to build new global geographies of production, trade and investment had collided with rising interest rates, inflated prices paid for assets, and problems of debt servicing.

These changing conditions forced many large companies to rethink their strategies. All of the corporate raiders examined in this chapter had begun to change direction by 1989-90. Industrial Equity Ltd. sold its food interests and its control over the largest food retailers, Woolworths Ltd., to Adelaide Steamships Ltd. producing a vertically-integrated structure from farm gate to supermarket. By 1990, Elders IXL Ltd. had also indicated that it would concentrate on the more secure cash flows from global agribusiness, food trading and brewing. It tried to sell a major share of its merchant banking arm to international financiers, while its own core management group sought to maintain control over the corporation by acquiring a controlling interest in Elders IXL Ltd., again largely with borrowed funds. International finance agencies quickly lowered the credit rating of Elders IXL because of the high level of debt incurred by its new parent.

The most dramatic changes have involved the Bond Corporation Ltd. which, by early 1990, hovered on the verge of financial collapse. The example illustrates the importance of cash flows to companies heavily in debt. After becoming beer market leader in New South Wales during the early 1980s, the brewer Tooheys Ltd. (acquired by Bond Corporation in 1985) lost market share rapidly and recorded losses in 1988 and 1989. Tooheys revealed that it had made a "loan" of more than A$1billion to its parent company which would have been swallowed by Bond Corporation Ltd. to help reduce its international debt. In a further attempt to avoid defaulting on interest payments, Bond Corporation Ltd. proposed selling a half share of the Australian brewing division, the only substantial generator of cash flow within its complex corporate structure, to New Zealand's largest brewer, Lion Nathan Ltd. Early in 1990 this strategy caused a consortium of the creditors of Bond

Brewing Ltd., fearing loss of control over its major productive assets, to have the brewing division placed in receivership.

Two things emerge clearly from the examples in this chapter. First, while the cash flows from food and beverage production attracted major internationalizing corporations during the 1980s, debt servicing and financial strategy began to outweigh decisions about production. Financial instability placed the performance and employment levels of some key production divisions in peril. Second, the most recent experiences in the Australian food industry demonstrate even more conclusively that theories of the internationalization of capital which emphasize the relocation of production fail to capture the complexity of global change in the industrial restructuring of the 1980s. Domestic market struggles, investment decisions and employment levels in food and beverage manufacturing are influenced increasingly by the global financial strategies of the major corporations.

ACKNOWLEDGMENTS

The authors would like to thank Gwen Keena, John Cleasby and Kathy Routh for assistance in completing this paper.

13. RESOURCE DEVELOPMENT AND THE EVOLUTION OF NEW ZEALAND FORESTRY COMPANIES

Richard LeHeron

The February 1987 C$385m takeover bid by Fletcher Challenge Limited (FCL) for British Columbia Forest Products Ltd. has made the company, following Canadian government approval, the world's second largest producer of both newsprint and pulp. A year before, in a different corner of the Pacific Rim, Carter Holt Harvey (CHH), another New Zealand forestry corporation, purchased into a Chilean joint venture with Maderas Prensudas Cholguan SA and Forestal Cholguan SA. On one level the takeovers represent the latest development in an extraordinary series of competitive moves by several New Zealand forestry companies, now radically restructured from the small companies that started internationalizing early in the 1980s. On another level the moves appear to be part of a new scale of co-ordination in at least the Pacific Rim segment of world softwood production. From a New Zealand perspective (either capital or labour) at least four axes are discernible in the rearticulation underway: New Zealand-wide restructuring of forestry activity, partly precipitated and accelerated by state action; reorganization of trans-Tasman forestry production; redefinition of Japanese and Chinese access to Pacific Rim wood fibre supplies; and changes in the nature of wood product consumption in the principal industrial markets of Japan and the United States.

To understand the emergence of New Zealand forestry companies, an approach is needed which situates evolving capital in changing contexts. This chapter gives a highly condensed and necessarily selective contextual account of that evolution, during an era of state-managed welfare capitalism in New Zealand and amidst a transition to a more open and less interventionist economy. The present chapter builds on work relating to the determinants of exotic afforestation and the emergence of a wood-based resource utilization industry in New Zealand (Le Heron, 1987a; Le Heron and Roche, 1985; Roche, 1987a). The emphasis is on identifying dimensions of structural possibility and constraint and revealed company action at different times. The New Zealand forestry experience is probably unique in world terms and perhaps paradigmatic in character, not because of internationalization per se, but because much of the forestry company offshore expansion has proceeded in parallel with a wholesale reduction of state intervention in the sector and the New Zealand economy as a whole.

Several issues of approach must be briefly mentioned. First, forestry production is conceptualized as capitalist (Le Heron, 1987b) and organizational and technological change are viewed as mostly consequent upon strategic moves by organizations to attain objectives, in the case of companies, satisfactory returns on investment in the face of intensifying competition (Le Heron and Roche, 1985; Scott, 1983). Industrial forestry thus involves the circulation of capital through the stages of tree planting, processing and reafforestation (to preserve utilization investments). The technological platform of industrial forestry (like much late twentieth century food and fibre production) rests on the application of science and technology to decrease the turnover time of capital and cut down biophysical risks. Under welfare capitalism industrial forestry became an identifiable conduit, often involving monopolistic companies, for the extraction of surplus value (Dargavel, Hobley and Kengen, 1985).

Second, although the geographical focus is New Zealand, the perspective adopted is international. This permits changes in New Zealand to be seen in relation to changes going on in other nations. Intepretative discusssion is informed by the political-economy literature that has re-examined theory about capitalist production in the light of the rise of global capital (Mandel, 1975; Cohen, 1981; Perrons, 1981; Castells, 1983; Gibson and Horvath, 1983; Susman and Schutz, 1983; Taylor and Thrift, 1983; Massey and Allen, 1984; Walton, 1984). In particular, the literature suggests the transition from welfare to global capitalism must be accompanied by restructuring of state-economy relations and general reorganization in the sphere of the economy and state, to establish new conditions for, and mechanisms to enable, accumulation. In concrete terms, restructuring is manifested in an upsurge in merger activity, involving penetration and divestment of capital across traditional industries, within and across countries and changes in the nature and degree of government intervention.

Third, conceptualizing forestry production in the general context of wider transition demands consideration of both changing structural determinants and organizational strategy to achieve or facilitate profitable production. Fortunately, periodization of structural relations, and relative autonomy of organizations helps in the identification of broad developments. Of special interest is the content of relative autonomy under particular structural relations and how organizations that have survived and grown in particular contexts respond as structural conditions affecting them alter. Fourth, obviously any contextual account cannot be all-encompassing. This chapter in fact focusses almost exclusively on aspects of private sector restructuring and explores few changes in the state arena. Moreover, the discussion privileges the investment and not the labour side of production. This is because the evidence indicates that labour questions have only been a minor influence in company decisions to internationalize.

13.1 THE STRUCTURAL CONTENT OF CAPITALIST PRODUCTION IN NEW ZEALAND

Land-based production in New Zealand has generally been highly constrained by the country's position in the capitalist periphery (Armstrong, 1978; Franklin, 1978; Taylor and Thrift, 1981) and the impact of a strong centralist state (Le Heron, 1987b; Welch, 1982). The formative years of industrial forestry in New Zealand were not, however, noticeably influenced by peripheral status. Rather, distinctive state-economy relations prevailing in the 1920s played a determining role in a local solution to pending wood supply problems. Indeed, the issue of New Zealand's structural relations with the rest of the world was temporarily shelved during the long era of welfare capitalism. Like many overseas counterparts, the New Zealand government increasingly regulated the exchange rate and the banking system, placed restrictions on foreign ownership and controlled imports, effectively isolating in New Zealand a mixture of domestic and foreign capital. These institutional arrangements defined the terms under which accumulation could occur and ultimately constrained the direction of national development. But which businesses succeeded or failed in this context depended on competitive pressures and particular advantages and disadvantages conferred by government policy. Structural questions regained prominence in New Zealand when the limits to welfare capitalism were reached. In 1984, the New Zealand government induced a general crisis by comprehensively revising the conditions facing investors. Less constrained than previously, New Zealand companies quickly capitalized on any advantages and attempted to internationalize their business. It is in these changing settings that the expansion of New Zealand's industrial forestry and the growth of the key forestry companies must be examined.

13.2 INDUSTRIAL FORESTRY IN NEW ZEALAND: CONDITIONS AND POSSIBILITIES

Industrial forestry in New Zealand can be traced to two significant developments in the 1920s. The first was the establishment of the New Zealand Forest Service (NZFS) which allayed public and investor fears that a timber shortage was probable in the mid-1960s. The NZFS catalogued the indigenous resource (Ellis, 1920), formulated policy to replenish the rapidly depleting native stands with exotic forests managed on a "permanent, perpetual and sustained basis" (Ellis, 1920, p.3); stocked nurseries and embarked on a planting program, commencing in 1923, that led to 170,000 ha of state exotic forest by 1935.

The second development, of similar magnitude, involved a move into afforestation by land development and bond holding companies (Roche, 1987). Promoters were enticed by publicity of a timber famine and NZFS pamphlets outlining silviculture techniques and possible returns from trees. Some 96,479 ha were planted between 1926 and 1933. Following a Commission of Inquiry, the largest bond holding company was reformed into a single proprietary company, New Zealand Forest Products (NZFP), which owned 59 percent of private plantings.

Even though harvesting was not scheduled before the 1950s, by the mid-1930s the issue of forest utilization had already become pressing. Tensions were evident between the NZFS and the afforestation companies and mainly pivoted around which organizations would, and under what terms, convert trees into industrial products. The NZFS largely accepted private forestry but was mindful of earlier crises in the indigenous sawmilling industry and sought "some measure of statutory control" (AJHR, 1926, C3, p. 8) over utilization companies. Early NZFS plans to enter utilization directly were abandoned by the 1940s and attention turned to the nature of timber sales contracts with selected private companies. At the same time, state policy gradually pushed the notion of New Zealand timber for New Zealanders

(Roche, 1987b). The government enforced an export ban from 1919 until 1928 and again for selected species from the late 1930s and severely restricted log and timber imports. The stage was now set, with a declining indigenous resource and fast maturing exotic plantations, for the growth of a protected utilization industry.

Several aspects of the New Zealand plantation program were to impinge on prospects for company growth. The New Zealand physical environment was extremely favourable for rapid tree growth. Trees could be grown and harvested within 30 years: a human generation. The scale of afforestation created large areas of relatively uniform trees. Science and technology refined the quality of the new wood resource. This technological platform conferred an unprecedented opportunity in New Zealand, and eventually internationally, for the processing of wood fibre. Most significantly, the form of state-private sector management of New Zealand's exotic forest resource proved to be an effective mechanism for surplus value extraction and retention by private companies. The details of these arrangements and company success amidst constraints are now considered.

13.3 FORESTRY COMPANY GROWTH: CONDITIONS AND CONSTRAINTS

The question of how to utilize the trees from the first national planting preoccupied both private and public sector planners for over 50 years. The principal lines of debate and subsequent state policy hinged on access to state forests, entry into the New Zealand market for paper and paper products, and the timing of new lines of manufacture. The simultaneous maturity of the "two national forests," one under company and the other under state ownership, posed a dilemma over how best to incorporate state wood into industrial production. Antagonism between NZFP and NZFS during the late 1930s virtually ensured entry of other capital (Healy, 1981). With no agreement after a decade of discussion on a joint NZFS-NZFP utilization proposal, resolution was reached when NZFS offered for tender the first wood from the State's Kaingaroa Forest. The proposed scheme was ambitious, featuring an international size sawmill, paper and newsprint mill, all using commercially unproven wood from *Pinus radiata*.

The scheme, although predicated on the perceived desirability of passing on much of the "unearned increment" from capital tied up in state forests to private industry, was in a sense the child of uncertainty over investor interest in an untried wood resource and the exceptionally large scale of funds needed to establish a state-of-the-art greenfield forest processing complex. To attract venture capital from overseas, the government packaged a comprehensive deal by committing itself to: sell at an unchanged price an annual supply of 651,28Om^3 of logs delivered to a nominated mill site and to contract deliveries at that rate for 25 years; a renewal of rights for two further periods of 25 years; sell a suitable site for an integrated mill; deliver an adequate supply of electricity; arrange for port construction and a rail link; provide rental houses and sections for employees; assist immigration of skilled workers; participate in capital structure; allow issue of debentures; and give overseas investors the right to transfer net profits. Significantly, in spite of worldwide publicity, only one offer was submitted, by a New Zealand company, Fletchers Ltd. The resulting company, Tasman Pulp and Paper Ltd (TPP), set up in 1954, had a shareholding of New Zealand government (37 percent), Arthur Reed Ltd. (25 percent), Fletchers (18 percent), Commonwealth Development Fund Corporation (9 percent) and the New Zealand public (9 percent).

Both NZFP and TPP were granted, via licences to import capital equipment, monopoly privileges in the New Zealand pulp and paper markets. The companies reached agreement over lines of production (Healy, 1981, p. 152), with TPP keeping to newsprint and kraft pulp and NZFP to kraft pulp, building boards and corrugated containers. Operationally, however, this still left the companies with a problem of disposing of production excess to domestic requirements. Both looked to Australia and drew heavily on earlier business connections; NZFP through bond company links and TPP via Reeds and then Bowaters. In the case of NZFP, arrangements were made with Australian Paper Machines (APM, now Amcor) to purchase quantities of pulp. TPP sold surplus newsprint to Australian Newsprint Manufacturers (ANM), a company dominated by two Australian newspaper publishers who consumed two thirds of Australian newsprint. The initial specialization spawned from state-company discussion and company marketing agreements influenced later company investment strategy and government interpretation of changing market opportunities. Steady mill expansion to cope with the peak volume of timber from state and private forests meant both companies could not expand — given the high interdependence of pulp and paper production — without generating additional output in the area nominally the specialist preserve of the other company. Initially exports were minor but by the mid-1970s NZFP exports amounted to 23 percent of total sales. For TPP exports were even more important, accounting for 58 percent of all sales.

Consistent with policy established in the late 1940s, the NZFS maintained a watching brief over the size and composition of the forest utilization industry. In 1969, against a backdrop of a mooted NZFP-TPP merger, stemming from the 1967 New Zealand Free Trade Agreement with Australia, the NZFS put the last uncommitted wood supplies in the Central North Island up for tender. Despite a joint NZFP-Fletcher-TPP tender, the contract was let to Pan Pac Ltd., a consortium of Carter Holt (CH) (60 percent), Kokusaku Pulp Industries Co. Ltd. (20 percent) and Oji Paper Co. (20 percent). For CH the contract shifted the company from secondary status (based on a sawmilling lineage) into a company with a rapidly growing export arm. By 1979, 50 percent of total product of all CH forestry units based on exotics (in dollar terms, 30 percent of sales) was exported. Like the two dominant companies in the industry, CH embarked on a program of expansion. A final allocation of state timber was made in 1976 to Winstone Samsung Industries, a venture between a New Zealand and a Korean firm.

The early growth of the New Zealand forestry companies was spectacular. NZFP quickly became the country's biggest company and TPP, trading in stable markets and benefitting from expanding output, made profits in the 1960s that "exceeded those of any other listed company" (Fletcher Holdings, 1987, p. 2). The companies were large employers: the pulp and paper industry and related products industry grew from a few hundred in 1950 to 9,220 employees by 1970. Growth, however, was a mixed blessing. The continued pulp and paper expansions of NZFP and TPP at the end of the 1960s were sufficiently large to outgrow the New Zealand term debt market. Offshore bank finance could be obtained only under stringent conditions, especially evidence of long term contracts at renegotiable prices. Agreements reached for the New Zealand market proved to be a millstone. TPP suffered from a 15 year forward sale agreement with the New Zealand Newspaper Publishers Association (signed 1968), with a 2 1/2 percent escalation clause. The effects of the new bases of funding were both cumulative and serious for the company. An inability to recover costs in New Zealand newsprint or from Australian pulp, successive devaluations which pushed up costs of servicing offshore loans, and extra construction costs and project delays resulted in a liquidity crisis in 1978. TPP was rescued by Fletchers, long dominant in the New Zealand construction scene, and which had only expanded into forestry in 1970, as a diversification move.

The late 1970s heralded the beginnings of an international strategy by the companies; a strategy that was in part conditioned by past investments but which also reflected a growing awareness of distinct limits to company growth in New Zealand, in forestry and in other sectors. To a degree, however, the content of company strategy had been prescribed for two decades by state promotion of export-led economic growth. Government encouraged a second national planting, through indicative planning. Company and NZFS plantings in different parts of the country, while satisfying the government objective of a twofold increase in exotic forest (so yielding increased foreign exchange earnings), were only roughly co-ordinated and yielded a continued concentration under dual state-company ownership, in the Central North Island.

A new dimension appeared in NZFS afforestation, with the management of state forests to a clearwood silviculture regime based on pruning (Elliott, 1982). In practice, both the NZFS and the companies, who did not widely adopt the new timber management regime, were less than successful in creating a forest resource with a wholly new character. A great advantage of the clearwood strategy was the ability (usually argued as undesirable, given investment to establish a clearwood regime) to downgrade the timber use from the high quality end to residual processing. A significant side effect of NZFS clearwood policy by the 1980s was an unwillingness by the NZFS to release timbers from forests on the verge of gaining their maximum value (when trees are 25-30 years old). The NZFS stance added to a growing timber shortage brought on by the gap between installed capacity (right across the forestry sector) and projected increases in wood volume through to 1990. As a result the companies were triggered into a reassessment of available wood supplies to feed existing mills. They had to confront the twin issues of how to utilize the newly planted forests and to establish alternative ways of fulfilling domestic and overseas obligations as they became increasingly conscious of the finite size of New Zealand as a market for forest products.

Thus by the 1980s a small set of New Zealand forestry companies controlled the bulk of industrial forestry. A second generation of exotic forests on an expanded scale ensured the long-term supply of timber. In this regard there was general agreement, though long range planning was plagued by uncertainty (Inter Regional Planning Group, 1983). In the short run, however, concerns about where to invest the steady profits obtained from milling first generation trees were paramount.

13.4 CHANGING CONTEXT AND CHANGING COMPANY STRATEGY

At the end of the 1970s, companies operating in New Zealand found it increasingly difficult to identify profitable large scale investments, in spite of concerted efforts by government to plan a series of "Think Big" projects, amongst them pulp mills to come on line in the 1990s. Indeed, Hugh Fletcher (1982, p. 6), Managing Director of FCL, by then the largest New Zealand company, (resulting from a merger of two companies with little forestry experience) expressed the general difficulty as not one of choosing between alternatives, but as a "project constraint." The incidence of company merger activity, a recognized strategy to ease competitive pressures, grew steadily from 1975 onwards (Fogelberg, 1984). The final phase of state intervention had kept the possibility of industrial forestry intact through assisted planting and subsidized forestry companies exports. Subsidies for the years 1975 to 1985, for example, amounted to at least NZ$362m for export market development, export incentive allowances and export credits and NZ$96.3m for forestry development assistance. But from the standpoint of the forestry companies the government's goal of greater export earnings was subordinate to their own survival and growth in a steadily more challenging business environment.

In contrast to most other New Zealand industrials those in forestry were both vertically integrated and had a working knowledge of overseas markets (Le Heron, 1980). Early attempts at restructuring in the sector, however, were aimed at reorganizing domestic production and markets. Fletchers made a bid in 1979 for CH, shortly after the Fletcher takeover of TPP management. This bid was abandoned when Fletcher Industries and Challenge Corporation merged in 1980 to form FCL. A year later the new company purchased an 84 percent holding in Crown Zellerbach Canada (renamed Crown Forests). The total cost of the acquisition was NZ$421m, financed by bank loans of NZ$232m and vendor finance of NZ$189m. FCL management enthusiastically acclaimed the prospects for "the expanded Group's activities in the world market for forest products [as] exciting. Strategic advantage will accrue from market presence" (FCL, 1981, p. 3). New Zealand development was seen in a new light, with the company stating that the decision to proceed with a No. 4 newsprint machine would be reviewed. Earnings contributions from Canadian operations between 1983-1986 exceeded company expectations (FCL, 1986).

NZP was relatively conservative in its growth strategy, opting for a combination of domestic and trans-Tasman realignment. Domestic takeovers had dominated NZFP strategy in the 1970s but the company was sensitive to its vulnerability for takeover. Diversification was adopted to "provide a source of increasing income in the shorter term" (NZFP, 1982, p. 10) until new forests matured. Parallelling this policy the company sought to strengthen APM and NZFP links through an exchange of directors. NZFP reacted defensively to the successful CH takeover of Alex Harvey Industries (so forming CHH), attempted to secure Odlins, a diversified New Zealand company with timber products distributorships; and purchased a 23.7 percent share of Watties Industries following a Wattie inspired Dominion Industries purchase of 24.9 percent of NZFP. CHH also entered Australian forestry, acquiring South Australian Perpetual Forests, which owned 27,000 ha of pine plantations, investing in a chip loading facility at Portland and forming Wood Exports Ltd. (78 percent CHH, 22 percent C Itoh Ltd, a Japanese trading company with substantial involvement in the supply of raw materials to the pulp and paper industry in Japan). These moves gave the CHH scope to use South Australian pulp (from a proposed 120,000 ton per year mill) to meet a substantial portion of its contracts to supply pulp in Japan, so freeing its New Zealand forest resources for the planned production of newsprint at Whirinaki.

It must be stressed that merger activity was both possible and completed when the economy was still closely managed by government. The significance of the 1984 economic "reforms" was to remove assistance, deregulate the economy, float the exchange rate and corporatize government trading departments. Companies already operating internationally were better prepared for the resulting pressures to internationalize and were in a strong position to exploit new domestic opportunities coming from delicensing and deregulation of the forestry products market. Post-1984, the companies kept up the momentum of overseas investment in forestry. CHH and eventually FCL moved into Chile. CHH's Chairman noted that only a global marketing strategy would allow New Zealand to successfully sell its forest products during the next three decades. CHH presence in Chile meant it might "Influence the processing and sale of radiata pine, and it would get a larger share of the world market for medium density fibre board" (Anonymous, 1986). NZFP, through its investment company Rada Corporation, bought into North Broken Hill Holdings, generating speculation in Australia that a rationalization of the Australian pulp industry was pending (Withers, 1986).

To fund the expansions, the companies made heavy borrowings, especially abroad so that the patterns of their long-term debt altered dramatically. FCL's NZ$849m debt is two thirds offshore, while NZFP records near 55 percent overseas debt. Despite the enormity, in New Zealand terms, of the CHH and FCL acquisitions and their potential Pacific Rim importance, the developments reflect in large measure medium term strategy. Immediate attention has focussed on the New Zealand scene where FCL has posted a takeover bid for NZFP. Prior to 1984, the notion of single private company ownership of most of the private forests and the forest utilization industry was a political anathema. However, in the changed business environment, neither the FCL move, nor the Rada counterattack, nor consequent NZFP talks with Amcor (probably implying Australian ownership) are unacceptable proposals. Substantial gains would accrue to the acquiring company. Should it be FCL, the New Zealand market would be open to rationalization, and the economics of running two forest products complexes located adjacent to one another at Kinleith under one ownership would be subject to scrutiny. An NZFP-Amcor merger, on the other hand, would enable the rationalization of New Zealand-Australia pulping capacity, with Australian mills emphasizing waste paper and hardwood and the New Zealand plants manufacturing higher grade white boards utilizing softwood.

13.5 RESOURCE UTILIZATION ISSUES

In New Zealand, industrial forestry of a distinctive kind was established and elaborated under the supportive arch of state intervention. The government mandate to see the forest resource effectively used was translated, through NZFS and company negotiation and legislation into low, long-term wood costs and a protected, restricted domestic trading environment. Operating primarily in the New Zealand market the companies collectively became a growth pole in the economy, expanding both output and employment on a steady basis for more than 30 years. Thus while the state defined the conditions of resource development under which the companies could conduct business, the companies enjoyed considerable autonomy providing they did not threaten dominant New Zealand ownership or jeopardize the availability of a full stream of forest products.

By reinvesting on a progressively expanded scale, the companies quickly shifted from minor status to become the largest industrials in New Zealand. The potential discontinuity in forestry production inherent in non-replacement of first generation trees was averted with the help of government export promotion programs which coincided with the years when massive replanting was needed to restock forests. Up to this time, the government and forestry company objectives appeared to be compatible. However, afforestation on an enlarged scale implied, at least for government, increased utilization. Whereas in the previous decades an assured market had existed for forest products, export-oriented production on a scale envisaged by government presented the rapidly internationalizing companies with a quite different investment environment. A principal effect was to reduce the time horizon for adequate returns on investment and to heighten a concern for market adaptability in general, rather than necessarily maximizing returns at the forest edge.

If the companies had not internalized this thinking by the mid -1970s, the view was well entrenched by the mid -1980s, when government, seeking to plan strategically the utilization of second generation forests, encountered a reluctance to make longer-term commitments. The undoubted success of early growth in forestry output and employment between 1950 and 1970, that is during an expansionary phase in government intervention, perhaps encouraged government policy-makers to stress the seeming inevitability of further growth contributions from the sector. The different origins of broad company objectives from those espoused by government had thus finally confounded the idea of ever-increasing forest resource development. Tangible evidence of company investment consistent with other objectives began to surface at about the same time that the intervention apparatus was being dismantled. Owning and controlling replanted forests of approximately 200,000 ha, the companies began to systematically invest overseas. The initial moves appeared ad hoc, but continuing purchases in Australia, Canada, Chile and Hong Kong soon suggested the companies were extending the scale of co-ordination over production and marketing. The Pacific Rim reach of New Zealand-based companies engaged in forestry opens up a wide variety of options for resource sourcing to meet market requirements.

13.6 CONCLUSION

The main argument in this chapter has been that the evolution of New Zealand forestry companies can be understood only in the context of wider industrial restructuring, in New Zealand and abroad. The chapter accepted the view that suggests the world capitalist economy is undergoing profound restructuring and is becoming more integrated than previously. Specifically, welfare capitalism, featuring monopoly capital, is being replaced by a new order of state-economy relations. The theory also suggests that the "world economy is the outcome of a whole series of countervailing forces operating at a whole series of scales" (Thrift, 1986, p. 13). The emergence of the New Zealand forest products industry as a small but important global factor is especially illustrative because of the path by which the companies grew. The capacity of the forestry companies to embark on internationalization had roots in a particular articulation of state-economy relations and it was the historically specific limits to these that helped pressure the companies to expand offshore.

Two further aspects are of significance. First, unlike the past, the future *may* see the companies exercising selectivity over which forests and which mills, in which countries, will be used to meet market demands. This clearly has implications for work forces and communities dependent on forestry. Second, the reader should not be left with just the impression that the New Zealand forestry companies are extraordinarily successful, though to date this seems to be so. Rather, the key message is that company survival and growth depends as much on the socially created context of operation, which delineates broad possibilities for investment, as it does on the enterprise of particular corporations.

ACKNOWLEDGMENTS

I am grateful to Mike Roche for discussions on the shape and content of the chapter.

TRADE AND INDUSTRIAL POLICY ISSUES

14. THE CANADIAN ECONOMY AND TRADE: A GEOGRAPHER'S ASSESSMENT OF THE MACDONALD ROYAL COMMISSION REPORT

Iain Wallace

The ambiguities of Canada's nationhood do not go away: they recycle. The tensions of accommodating the interests and sensibilities of two founding peoples and of significantly different regional cultures and economies, and those that arise from living next door to the world's still dominant power combine to ensure that the direction and legitimacy of national policies are only rarely not active sources of domestic political friction. The particular focus of contention shifts over time, but the "resolutions" which allow this redirection of energy are never final. Neither the rejection of separation from the rest of Canada by the electorate of Quebec in 1980, nor the acceptance by the Government of Canada of a Free Trade Agreement with the United States in 1988 can be regarded as permanently solving the dilemmas of Canadian existence. Such is the context for this review of the country's interregional and international economic relations.

Publication of the Report of the (Macdonald) Royal Commission on the Economic Union and Development Prospects for Canada in 1985 marked one of those significant turning points in the orientation of national attention (Canada, 1985a). It followed a decade of introspection, spent securing the political integrity of the Canadian nation against the threatened departure of an independent Quebec. The Report's principal focus turned to Canada's place in the wider world, specifically its economic well-being in a menacing global trading environment. The Commission's conclusion, that contemporary realities point to free(r) trade with the United States as the most promising route to continued prosperity, was received sympathetically by the Conservative federal administration. But what would this, or should this, involve, and what would be the domestic consequences (not least the spatial implications) of bilateral commitment to pursue it? The Report was somewhat ambiguous on these questions. Indeed, it has been legitimately characterized as a "jigsaw puzzle" (Raynauld, 1986) of diverse and sometimes contradictory observations. Certainly, its 1,620 pages of text cry out for tighter editing.

Despite the congruence of the Report's free trade philosophy with contemporary political conservatism and the traditional ethos of Canada's resource-based hinterland economies, its espousal came at a time of severe and potentially lengthy depression in world commodity markets, underlining the continued vulnerability of Atlantic and western Canada. Yet the greater economic diversification enjoyed by Ontario and Quebec, owing much to a century of tariff protection, could be seen as equally insecure in a BFT environment. Nor was it clear that the United States stood to gain benefits from freer bilateral trade adequate to justify the political costs to many of its legislators. So it remained an open question whether bilateral free trade (BFT) offered Canada a genuine or attractive opportunity to escape the problems associated with the contemporary restructuring of the global economy. A critical review of the case developed in the Report, particularly its treatment of the issues confronting the evolution of the national space economy, seems an appropriate means, in light of the subsequent Free Trade Agreement, of introducing Canadian perspectives in this comparative volume.

14.1 THE STORMY SEVENTIES

A number of related developments subjected federal-provincial relations in Canada to severe stress during the 1970s and led to a neglect of international issues. They included an intensification of regional and cultural alienation, growth in the ambition and competence of provincial administrations to act independently (at Ottawa's expense), divisive shifts in the geography of resource rents, and the geographical pattern of electoral representation during the period. Even where the sources of friction lay in events outside the country, notably the OPEC-induced energy price rises, the consequences were invariably felt in Canada as profoundly disruptive *domestic* issues.

Harris (1982) has characterized Canada's discontinuous ecumene as an archipelago, whose settlement history and political evolution has produced a profoundly regionalized nation. That there should be distinctive regional interests, and tensions between them, within the Canadian state is therefore unremarkable, and the consolidation of national sentiment and identity over the past 120 years is no mean achievement. Nevertheless, the Macdonald Commission Report (hereinafter cited as Canada, 1985a, vol., page) recognizes that the tensions of Canadian federalism in the 1970s were so acute precisely because "the present design of our national institutions gives remarkably little attention to regional interests" (Canada, 1985a, 3, p. 72). The post-Confederation reliance on a regionally representative federal cabinet to ensure each part of the country a voice in national affairs breaks down if national political parties fail to win support across the country. This they conspicuously did between 1972 and 1984, when the Liberals lost the West and the Conservatives, Quebec.

The absence, in Ottawa, of effective representation for major regions of the country accelerated the emergence of provincial governments as the voice of disenfranchised regional interests. This move was fed by other powerful forces. Following Ontario's nineteenth century example of "the politics of development" (Nelles, 1974), other jurisdictions embarked on strategies of "province-building" through resource exploitation and industrial diversification, to the extent that their growing economies and bureaucratic capabilities allowed. What Quebec and British Columbia achieved through hydro-power in the 1960s, Saskatchewan sought to attain through potash and uranium, and Alberta, Newfoundland and Nova Scotia through hydrocarbon resources (or the search for them) in the 1970s. This activity brought the provinces into conflict with Ottawa in two areas. First, it frequently had international implications (for example, on the Columbia River; expropriation of US-owned potash mines) which countered federal policy and impinged on its sovereignty. Second, after 1972 it increasingly involved a federal-provincial battle over the division of greatly increased natural resource rents (an interprovincial battle in the case of Churchill Falls hydro-electricity). Moreover, the resource-based hinterlands became more articulate in their disaffection with policies seen as perpetuating their dependent status, such as discriminatory freight rates and the financial control exercised by central Canadian banks.

The place of Quebec within the Canadian state and of francophones within Canadian society was the most decisive question needing "resolution" in the 1970s. The election of an avowedly separatist Parti Quebecois provincial government in 1976, leading to the 1980 referendum on whether an independent Quebec should seek "sovereignty association" with the rest of Canada, had major economic repercussions. First, it fostered a "balance sheet" evaluation of Confederation at precisely the time that Ottawa felt impelled (for reasons explained below) to require Alberta to forego the full market value of its hydrocarbon resources in the interests of minimizing regional disparities. Second, it made the health of the Quebec economy a disproportionately prominent concern of federal policy, which, given the difficulties facing the province's low-productivity agriculture and labour-intensive manufacturing industries, fostered job-conserving protective measures at the expense of strategic restructuring. Third, it established unambiguously the cultural re-evaluation of economic performance associated with Quebec's "quiet revolution." The vision of becoming "maitres chez nous" (masters in our own house) among a rising generation of technocratically proficient Francophones extended both to harnessing the (provincial) state to this purpose and to supplanting the Anglophone elite in control of the Montreal business community. The immediate geographical result was to trigger a number of corporate relocations, primarily of headquarter offices, from Montreal to Ontario and to depress the level of industrial and commercial development in the province.

The rapid escalation of commodity prices in the early 1970s, which affected grain and metals as well as oil, gave the western provinces their first real opportunity to capture significant non-renewable resource rents. During this critical period, Manitoba, Saskatchewan and British Columbia had NDP governments that were predisposed to increase the "people's" return from constitutional ownership of the provincial resource base and who significantly increased mining royalties. This brought them into conflict both with the industry, which argued the fleeting nature of "super profits" and the need for taxation regimes to recognize its cyclical pattern of earnings, and also with the federal government, insofar as (moderate) provincial royalties had traditionally been a deductable expense with respect to federal taxation (Burns, 1976). Resolution of this issue became submerged, however, in the much more critical contest over the treatment of oil revenues.

The revolution in world energy prices caused such friction in Canada because the new rents accrued almost entirely to one province (Alberta), creating rapid and destabilizing shifts in regional prosperity. They also, thereby, imposed massive obligations on the federal government to provide increased equalization payments to the other provinces and to assure security of oil supply to import-dependent eastern Canada. The federal decision not to let Canadian oil prices move immediately to the new world price was defended on the grounds of shielding the weaker economies of Quebec and Atlantic Canada from a major additional burden and of giving Canadian manufacturers a competitive edge in international markets. But underlying this policy was the impossibility of accommodating Alberta's full windfall within the established federal mechanisms of equalizing provincial fiscal capacity. Naturally, Ottawa's moves to put a lid on Alberta's hydrocarbon income and to gain a larger proportion of the resource rent than hitherto were interpreted by many observers as the latest and most blatant attempt ever to favour central Canada at the expense of the West.

The Canadian public was generally oblivious of the fiscal pressure building up on Ottawa and even unaware that equalization payments are a charge upon the federal government rather than a sharing of revenues by richer provinces with poorer. The fact that by 1977-78 Ontario would have qualified to receive equalization payments had not the formula been amended is a measure of Ottawa's dilemma, and of the magnitude of the transformation of the geography of public sector income in Canada that had been brought about by OPEC. The generous inducements to hydrocarbon exploration in frontier regions which were part of the controversial National Energy Policy of 1980 were Ottawa's bid to generate rents which would not have to be shared with the provinces or enter into equalization calculations because they were derived from federal "Canada Lands."

Compared to the challenges of Quebec and Albertan oil, the issue of whether Canada needed an industrial policy — and if so of what sort — was generally viewed as less pressing. The federal Liberals under Trudeau espoused a nationalist policy with respect to ownership, given focus with the creation of the Foreign Investment Review Agency (FIRA) and subsequently with that of Petro-Canada to assert a major domestic presence in the oil industry. The ardour with which the policy was pursued was moderated, however, by its general unpopularity with the business sector and by continued interest in attracting large employers (invariably foreign transnationals) to depressed regions. It was left to geographers at the Science Council of Canada to raise pointed questions about the nation's industrial performance against the background of a restructuring global economy (Britton and Gilmour, 1978), but their advocacy of a policy of technological sovereignty as a basis for stimulating competitive domestic manufacturing failed to elicit support.

14.2 CANADA IN THE GLOBAL ECONOMY OF THE 1980s

The Macdonald Commission was appointed in November 1982 in the aftermath of a turbulent decade, and in the trough of the worst recession to hit the

Canadian economy for 50 years. Its broad terms of reference were predicated, in part, on the recognition that an increasingly competitive and interdependent world necessitated greater understanding of the aspirations of the regions of Canada, greater co-ordination between actions of governments in Canada, and greater support for the Canadian economic union (Canada, 1985a, 3, p. 561). The following analysis of the Report is organized around international, industrial (structural) and regional issues. The first group addresses Canada's place in the global economy of the mid-1980s and beyond.

The Commission was disturbed by the implications for Canada of "the trend towards regionalism in the world trading system...[given that it] is one of the few major industrial countries lacking free access to a market of over 100 million people" (Canada, 1985a, 1, p. 245). The impact that expansion of the European Economic Community has had on Canadian trade gives substance to this concern, as does the resultant "tendency for issues of world trade policy to be settled in trilateral negotiations involving the United States, Japan, and the...Community" (ibid). These developments are claimed to mean that Canadian manufacturers, lacking the advantages of specialization and economies of scale derived from supplying a large "domestic" market are less competitive internationally; and that Canadian exporters are likely to be caught in the crossfire of trade disputes between the principal actors (as is currently the case with respect to grain sales). Canada's attempt in the early 1970s to establish a form of economic relationship with the European Community which would counterbalance its heavy and, to nationalists, resented dependence on the United States came to nothing. Bilateral ties between Canada and Japan are not viewed as a feasible option at the current time. Hence the Commission concluded that "[g]iven Canada's increasingly isolated position in the international economy at large, it is difficult...to conceive of a genuinely effective alternative trade policy [to freer trade with the United States]" (Canada, 1985a, 1, p. 350).

The advocacy of BFT with the United States by a Commission supportive of Canada's longstanding commitment to multilateral trade liberalization within the framework of GATT points to the country's lack of viable options. Canada is the only major industrial nation for whom liberalized trade has become less diversified trade. Between 1954 and 1984, the US share of Canadian imports stayed at 72 percent, while its share of Canadian exports rose from 60 to 76 percent. Canadian trade policy cannot but give priority to the health of this overwhelmingly dominant bilateral exchange. This focus became particularly crucial in the early 1980s, with the proliferation of non-tariff barriers (NTBs) threatening access to the US market for both manufactured goods and resource products, and hence, potentially, all regions of Canada. Secure access to the continental market was also seen as the environment which will allow the scale and specialization weaknesses of Canadian manufacturing industry to be finally overcome. Last, but not least, the economic benefits flowing to Canada from BFT were forecast to yield domestic political dividends: "the strengthening of national unity and the removal of one of the most persistent and corrosive sources of regional alienation" (Canada, 1985a, 1, p. 357).

Of Canada's other trading links, those with the Pacific Rim receive greatest attention in the Report. They constitute the second largest set of commodity flows, having overtaken those between Canada and western Europe, but still account for only 10 percent of total Canadian trade. The Commission recognized that the Pacific region is "now the most dynamic in the world economy," but that Canada "is a relative latecomer" in a highly competitive market, whose Pacific trade, "while buoyant, has not been booming" (Canada, 1985a, 1, pp. 251/4/2). The Report acknowledged that Canadian indecisiveness towards the region is as much due to the domestic dilemmas which trade relations expose as to the frustration experienced in trying to export end-products to Japan. Indonesia has pointedly addressed Ottawa's ultimately untenable stance of trying to sell capital goods and professional services to NICs while simultaneously actively restricting the entry of their exports into Canada. With respect to Australia and New Zealand, it was assumed that, on the basis of their own recent commitment to freer bilateral trade, they would view "with some understanding" Canada's pursuit of an agreement with the United States (Canada, 1985a, 1, p. 261).

How credible was the Report's reading of Canada's international options? Its basic assessment of the country's limited leverage in the global economy is correct, and its modest expectations of significant advance towards further multilateral liberalization in the near future appear to be justified. Undoubtedly the most serious and pressing current threat to Canadian trade was US protectionism, embodied in NTBs rather than in tariffs. In April 1986 there were 300 protectionist bills before the US Congress, potentially affecting more than C$6 billion worth of Canadian exports and 146,000 jobs (Anderson, 1986, p. B2). To that extent, the recommended exploration of an agreement which better guaranteed access of Canadian goods to the US market was an initiative which no federal government could afford to ignore. On the other hand, the Report made no serious attempt (merely two pages of text) to document the nature and degree of support which might be enlisted in the United States in pursuit of BFT, nor, perhaps more critically, the extent of the opposition which it was likely to encounter. The Report's lack of discussion of a fall-back position is therefore a weakness, but the Commission appeared to see the preparatory *domestic* adjustments required for BFT as capable of significantly improving the performance of the Canadian economy, whether or not the Americans co-operated.

14.3 THE FUTURE OF CANADIAN INDUSTRY

The questions addressed in this section focus on the structure and performance of Canadian industry. The Report treated them within the overall context of its support for BFT with the United States. Does Canada's future continue to rest with its resource-based industries? Does the country need an "industrial policy" supportive of high technology manufacturing? Could Canadian producers compete effectively in a continental market under BFT? The Commission's discussion of these and related issues is rarely straightforward (Shearer, 1986), and subsequent public debate did little to clarify things. Allusions to the Canada-United States Auto Pact of 1965 were a particular source of ambiguity, for the benefits which are seen to have flowed from it, and which it was desired to foster in other branches of Canadian industry, resulted from a *trilateral* agreement (Canada, the United States and the Big Four auto makers) of narrowly *sectoral* scope. The much more comprehensive Free Trade Agreement in no sense replicates this structure.

The Report reached no consistent conclusion concerning two related questions. Does Canada require some form of interventionist industrial policy (including regional industrial policy); and if it does, is this compatible with a BFT agreement with the United States that satisfies American concepts of "fair" trade? On the one hand, the Commission argued that, "How we *produce* our economic goods and services is not ground for argument in the Canada-US free-trade debate. The point at issue is the manner in which these goods and services are *traded* "(Canada, 1985a, 1, p. 354, italics in the original). This is claimed in the context of denying that BFT would involve the dissolution of a distinctive Canadian culture and political economy. Elsewhere, however, the Report acknowledged that, "Since Canada's industrial and trade stance are so closely intertwined, a commitment to freer trade must be accompanied by a commitment to 'level the playing field' " (Canada, 1985a, 2, p. 202). This was argued to mean that Canadian policies, concerning tax structures for example, must be "sufficiently similar" to those of the United States to allow Canadian firms to be truly competitive south of the border. But disputes such as that over the nature of British Columbia's stumpage charges (a factor in the softwood lumber confrontation of 1986) soon gave substance to an alternative interpretation: that to satisfy the United States that competition under BFT is "fair" Canadian policies would be forced to match those of its neighbour much more closely, threatening the maintenance of a distinctive Canadian political economy.

The Commission advocated industrial policies which are "more harmonious with market forces than past practice has made them" (Canada, 1985a, 1, p. 52). Neither the path of tariff-protected import substitution on which Canada embarked with the National Policy of 1879 (leading to the growth of "miniature replica" branch plant industrialization in the 1950s), nor the interventionist approach of promoting high technology "winners" advocated by the Science Council of Canada found favour. Rather, "the principle of comparative advantage must be the primary determinant of our position in the international division of labour" (Canada, 1985a, 1, p. 63). Yet the Report acknowledged that this "doctrine ... fails to account adequately for a good deal of the international trade that is currently going on around the globe" (Canada, 1985, 1, p.157). The idea of an industrial policy designed to engineer long-term comparative advantage was seen to be "intriguing", but at the same time it "can prove highly disruptive to traditional concepts of 'fair trade..."(Canada, 1985a, 1, p.194).

So, should Canada trust its future to the market, or should it join the ranks of the "engineers"? The Report found "[t]he debate over the appropriate role of government in industrial policy...inconclusive" (Canada, 1985a, 1, p. 159). It nevertheless opted for the efficiency of the (enlarged continental) market over the potential for resource misallocation inherent in a targeted "industrial policy." The Commission's principal interest was clearly in the future of Canadian manufacturing: it ignored services (anticipating difficulties in including them within a BFT agreement) and, despite devoting a hundred pages to the natural resource sectors, it was strangely unspecific as to their prospects (see Shearer, 1986). It recognized that they will play an important but relatively smaller role in the future than they have in the past. It acknowledges that the energy megaproject focus of the federal government's 1981 "Economic Statement" was based on the erroneous assumption that "a permanent improvement had taken place in the terms of trade for natural resources in relation to those for manufactured goods" (Canada, 1985a, 2, p.142). And it voiced the now widely accepted recognition that "concern must be shifted from the issue of how quickly natural resources can be exploited to how well they can be managed" (Carmichael, Dobson and Lipsey, 1986, p. 25).

The Commission was pleasantly surprised by its sectoral studies of Canadian manufacturing performance, given its expectation that they would display "consistent weakness" as a result of tariff protection (Canada, 1985a, 1, p. 337). Its research confirmed Matthews' (1985) study for the Economic Council of Canada in concluding that over the period 1960-1980 Canadian manufacturing made a relatively successful adjustment to the changing global trading environment. Canada's growing deficit in high-technology goods, which figured prominently in Britton and Gilmour's 1978 study, is shown to have remained fairly constant as a proportion of total trade in these products, matching a stable level of "implicit self-sufficiency" for the manufacturing sector as a whole. Yet the irony of the Report's market-led, BFT-stimulated industrial policy is that Canadian industry is to be made more efficient by a closer integration with an increasingly protectionist and globally uncompetitive neighbour. This generates a number of concerns. The economies of scale and specialization argument is most appropriate to the manufacture of mature, standardized products for which the NICs have a clear comparative advantage over a growing range of goods. In contrast, the rapidly expanding capabilities of computer-integrated manufacturing, favouring smaller production plants and giving greater flexibility of output and profitability at lower break even volumes (Schoenberger, 1988) represent the trend in technological innovation from which Canadian producers stand particularly to gain. Moreover, the Commission argued that "it would be a mistake to view foreign [especially US] ownership as a negative factor" (Canada, 1985a, 1, p. 244), not least because of the benefits that can accrue from access to the parent's technology. But in seeking closer ties with the United States, Canadian manufacturers are in danger of preoccupying themselves with a market that is no longer (as it was in the 1950s) a reliable source of state-of-the-art technology. Perhaps the Commission's assurance of the effectiveness of BFT as a solution to the challenges facing Canadian manufacturing in a dynamic global trading environment can be put down to its neoclassical faith in the benefits of "access to the expanded unrestricted market and ... economies of scale." It was "impressed" by the consistency "over the last 30 years" of this "mainstream opinion" among Canadian economists (Canada, 1985a, 1, p. 327).

14.4 THE FUTURE OF CANADA'S REGIONS

It is entirely consistent with the commitment expressed above that on page 200 of Volume Three the Commission acknowledged that, "[t]o this point in our Report...[o]ur discussion has proceeded as if all economic activity took place at one geographic point." The Report is well aware that Canada is a country of regions and is concerned, as noted earlier, that they are not well served by existing political institutions: hence its recommendation that there be a regionally representative elected Senate. But when it comes to analysing economic regionalization, the Commission was less sure of how to proceed: "[f]ew of the issues in [its] mandate have proved more perplexing than regional development...Relatively little is known about how and why regional economies grow" (Canada, 1985a, 3, p. 198). It finds that the "traditional heartland-hinterland view of Canadian trade policy is now being challenged from several quarters" (Canada, 1985, 1, p. 331). Indeed,

> If factors of production are mobile between regions...as they now appear to be, how can it be determined whether a given region gains or loses from a change in trade policies? Some of the residents will be affected, but if workers move in and out of regions in response to changes in trade policy, the size of the region is not fixed, and interregional effects of trade policies become hard to 'nail down' (ibid).

Whatever the theoretical difficulties of regional analysis, the reality of Canadian political economy is that provincial governments treat their territories as regions, have considerable power to influence economic activity within their borders, and have developed a range of policies that impinge directly on interprovincial trade. Provinces occupy specific parts of the national space-economy and have a continuing institutional identity despite the considerable flux of interregional migration. They constitute one, if far from the only, meaningful framework for regional analysis, and it is surprising that a Commission enjoined "to achieve greater understanding of the aspirations of the regions of Canada... and greater support for the Canadian economic union" (Canada, 1985a, 3, p. 561) did not analyse more incisively the spatial dynamics of provincial economic performance.

TABLE 14.1

TRADE BALANCES ON GOODS AND SERVICES BY PROVINCE, 1979

	(000 $)		
	Balance with Other Provinces	Balance on External Trade	Overall Balance
Newfoundland	-1,174,790	878,400	-296,390
P.E.I.	-275,790	24,140	-251,650
Nova Scotia	-850,880	-871,870	-1,722,750
New Brunswick	-1,264,960	173,380	-1,091,580
Quebec	218,660	-4,294,840	-4,076,180
Ontario	7,463,330	-7,357,420	-205,910
Manitoba	108,350	-55,790	52,560
Saskatchewan	-1,862,010	2,087,500	255,490
Alberta	1,710,800	2,984,310	4,695,110
British Columbia	-3,683,430	4,114,790	431,360
Yukon and N.W.T.	-389,280	343,790	-45,490

Source: Calculated from Statistics Canada, Input-Output Division, Interprovincial Trade Flow Data, 1979.

Source: Canada, 1985a, 3, Table 22.5.

The Report's aggregate analysis of interprovincial trade flows (Canada, 1985a, 3, pp.101-109) indicates the diversity of trade orientation among the different regions (Table 14.1). These data do not seriously challenge the heartland-hinterland interpretation of the Canadian economy described in Chapter 4. Only central Canada (overwhelmingly Ontario, and including Manitoba because of a positive balance on services) and Alberta (because of fuels) are net interprovincial exporters, whereas positive balances on external trade are registered throughout the hinterland (except for Manitoba and Nova Scotia — where oil imports are probably responsible). The Commission also noted that interprovincial transactions constitute only about one fifth of national output, and it did not believe that this proportion is seriously diminished by internal barriers to trade. Nevertheless, it regarded the reduction of NTBs to trade among the provinces of Canada, which would necessarily be hastened by BFT with the United States, as highly desirable.

The Commission focussed on equalization payments and regional development policies as the two principal mechanisms whereby federal governments have attempted to alleviate "the persistent regional disparities in the Canadian economy" (Canada, 1985a, 3, p. 180). It reviewed the strains to which equalization was subjected in the 1970s by the growth of Alberta's energy revenue, and noted the continuing arbitrariness in its calculation, a potential source of renewed federal-provincial dispute. Nevertheless, the success of these payments — together with other federal transfers — in equalizing fiscal capacity across the country has been quite substantial. Interprovincial disparities in income and unemployment rates have not proved so responsive to public policy, although the starkness of comparisons of market income per capita (with a range of 53.8 percent to 114.1 percent of the national average between Newfoundland and Alberta) is softened on the basis of real personal disposable income per household (81.5 to 111.4 between Newfoundland and Saskatchewan) (Canada, 1985a, 3, p. 202: 1981 data).

The Commission's review of federal regional development policy and of its institutional framework was brief and drew on previous studies, notably the Economic Council of Canada (1977). The continuous bureaucratic restructuring and near-demise of this field of activity in recent years have been as much symptoms as causes of the difficulties of making tangible progress towards program objectives (Cannon, 1989; McLoughlin and Cannon, Chapter 17). The basic fact is that "there is...less variation in per capita incomes across regions now than there is among individuals within provinces" (Canada, 1985a, 3, p. 216). This led the Commission to argue that the federal government should quit the field of direct regional job creation and leave "place prosperity" to the provinces, where such a concern is "understandable and defensible" (Canada, 1985a, 3, p. 219). The federal role should be to ensure "person prosperity" through its nation-wide transfer mechanisms and through the enhanced support to occupational adjustment which the Report advocates. The Commission nevertheless believed that "the total federal commitment to regional development, [delivered through federal-provincial channels], should increase significantly over the next few years" (Canada, 1985a, 3, p. 220). Yet the strong American reaction to past regional incentives, such as those given to the Michelin plants in Nova Scotia, should have alerted the Commission to the incompatibility of such policies with the "level playing field" associated with BFT.

The Report devoted very little space to urban Canada as such, despite acknowledging that "local government...is big government," and that half the national population is now concentrated in the 10 largest cities (Canada, 1985a, 3, p. 373). One is left to assume that the preoccupation with the problems of urban growth which characterized the early 1970s (and led to the creation of the short-lived federal Ministry of State for Urban Affairs) was not unproductive, for "Commissioners are conscious that Canadians have been doing something right at the level of the local community...[and have] developed some of the most liveable cities in the world" (Canada, 1985a, 3, p. 374). The Report argued that, because of their awareness of local issues, "in general, local governments can most effectively establish and carry out urban policy", but it noted their limited constitutional powers to "plan in a positive way" (Canada, 1985a, 3, pp. 383/375). Because economic development is expected to become "still more urban," and the quality of urban life therefore an increasingly important "economic asset," the Commission favoured transferring greater development responsibility and more fiscal capacity to municipal governments, at the expense of the provinces (Canada, 1985, 3, pp. 383/408). Although there is a significant degree of justification for this positive review, it fails to register the polarization of society which is becoming increasingly evident in Canada's largest cities, where food banks have become institutionalized and an acute shortage of low-income housing is as prevalent as the construction of the kinds of post-industrial monuments to the gods of work and play discussed by Ley and Mills in Chapter 7.

14.5 CANADIAN CHALLENGES: CANADIAN CAPABILITIES?

The foregoing assessment of the Macdonald Royal Commission Report indicates the limited degree to which geographical insights and a sensitivity to spatial distributions informed its analysis. Predictably, the federal government's subsequent negotiation of a Free Trade Agreement with the United States generated intense public debate and revealed significant regional differences in support for BFT. The issue so dominated the 1988 election that the vote almost served as a national referendum. The unanticipated element was the strong endorsement of the Agreement in Quebec, despite the provincial economy's potential adjustment problems, which contrasted markedly to sentiment in Ontario and some of the hinterland provinces (Frizzell et al., 1989). Analysis of the spatial impact of the Free Trade Agreement, which took effect in January 1989, is currently premature. But events since 1988 have confirmed that the Canadian space-economy remains exposed to the impacts of the ongoing global restructuring of economic activity, with or without the Free Trade Agreement.

The ability of Canadians to respond creatively to economic changes in the wider world, which is essentially what the Macdonald Commission was intended to enhance, will continue to depend heavily on how well they manage their domestic affairs. In this respect, the 1990s have not dawned brightly. Despite sustained economic expansion since the recession of the early 1980s, the federal government has failed to shrink its annual deficits significantly, so interest payments loom larger than strategic investment in its spending. Capital spending by industry has been healthy, but this has shown Canada's import-dependence in technology-intensive goods to be no less than it was a decade ago (Britton and Gilmour, 1978). Most depressingly, but perhaps entirely predictably in the light of my opening paragraph, constitutional and linguistic tensions, focussed around the now apparently failed Meech Lake Accord, have returned to monopolize the emotions and attention of the nation in an atmosphere calculated to undermine rather than to promote constructive domestic accommodation to a changing world.

FIGURE 15.1

SIGNIFICANT TRADE FLOWS AMONG PACIFIC ECONOMIES (1979)

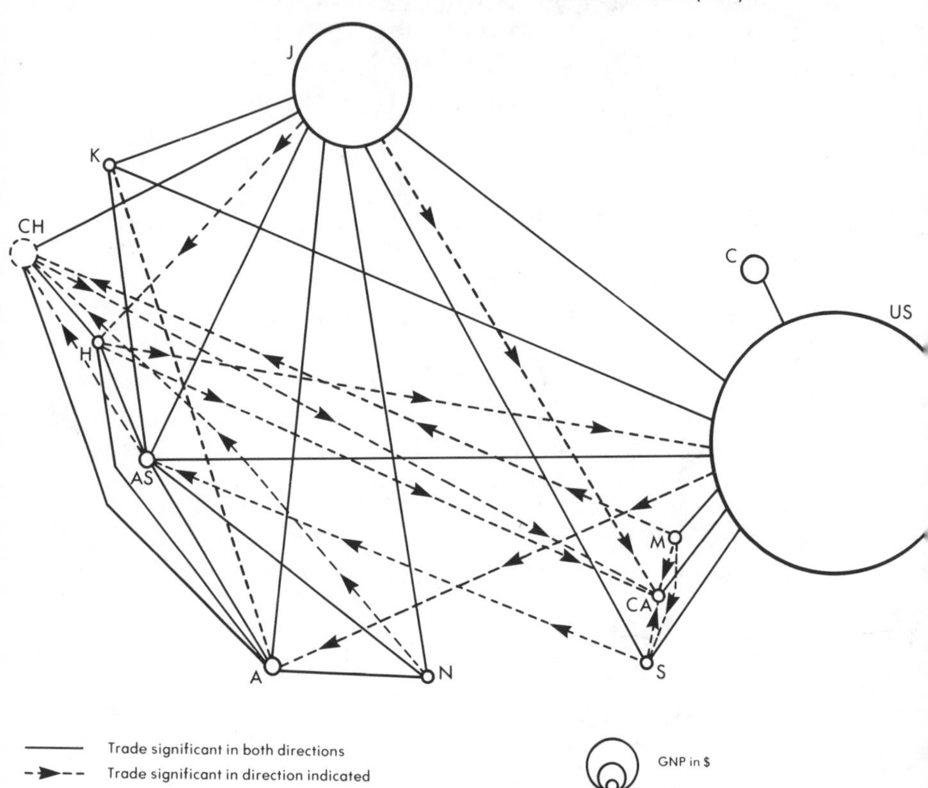

C = Canada
US = United States
M = Mexico
CA = Central America (El Salvador, Guatamala, Costa Rica, Nicaragua, Panama, Honduras)
S = South America (Columbia, Ecuador, Peru, Chile)
N = New Zealand
A = Australia
AS = ASEAN (Malaysia, Singapore, Thailand, Indonesia, Philippines)
H = Hong Kong
CH = China
K = S. Korea
J = Japan

15. CANADA'S TRADE AND INVESTMENT LINKS IN THE PACIFIC REGION

Roger Hayter

This chapter examines Canada's search for a more appropriate role in the world trade of industrial goods, particularly with respect to participation in the Pacific region. By way of context, the evolution of Canadian thinking regarding trade policy is briefly outlined. Next, Canada's trade relations with Pacific economies since 1971 are noted and then related to international investment patterns. The last part of the chapter focusses on the relationships between trade and investment. From a policy perspective, it is suggested that more effective participation by Canada in the Pacific region, or in the global economy at large, will require major initiatives in the public and private sector to stimulate the international growth of firms by direct foreign investment as well as by exporting.

15.1 CANADIAN TRADE POLICY: THE FIRST, SECOND AND THIRD OPTIONS

In terms of the political and economic relations among Pacific countries, Canada remains very much an outsider. Government and business approaches towards greater involvement in the region are — with exceptions — limited, unco-ordinated and passive. Indeed, according to Ross (1982, p. 45), attempts by the federal government to participate more formally in the evolving multilateral (and bilateral) relationships in the region, can be characterized as a "prudential evasion of formal commitments."

Some observers would object to this assessment (e.g. Delworth, 1982). After all, Canada's trade, investment and foreign aid ties with Pacific countries have expanded rapidly since 1950 within the context of sharpening political awareness of the role of the Pacific region in the global economy. Moreover, pronouncements by members of the federal government, beginning in 1970, for the Third Option Policy ostensibly represented a desire to reduce the impact of the US on Canada by increasing domestic control of the economy and by trade diversification, particularly with respect to the European Economic Community (EEC) and Pacific countries. Thus, to help repatriate the economy the government established the Foreign Investment Review Agency (FIRA) in 1973 and the National Energy Program (NEP) in 1979 while numerous high profile attempts were made to stimulate exports to Pacific countries including exports of manufactured end products. As formal policies, the First Option, to maintain existing relations with the US, and the Second Option which proposed closer integration and co-operation with the US, were not favoured by the federal government at the time.

Canada's trade with Pacific countries did increase relatively and absolutely during the 1970s. This increase, however, was associated with a relative decline in Canada's EEC trade. In trade terms, Canada became even more tied to the US during this period, a trend consistent with the rejected Second Option Policy. In effect, the Government was unable to convince or support the business sector to undertake a sufficiently strong commitment around the Pacific to offset Canada's growing reliance on the US. Indeed, to an important degree, Canada's Pacific trade has been initiated by Asian, especially Japanese business (Langdon, 1983). In the context of trade, the Third Option was an illusion, or as Langdon puts it, never more than a "formulation" (Langdon, 1983, p. 78). Moreover, other key elements of the Third Option, including the NEP and FIRA, have been dismantled. It also appears that the stimulus for the recent opening of free trade talks with the US, which as Wallace notes in the previous chapter was strongly advocated by the Macdonald Royal Commission, was predicated on increasing Canada's continental ties and therefore to explicitly take on the thrust of the Second Option.

For Ross (1982), formal and extensive political commitments by Canada in the Pacific, for example, with respect to defense matters or the creation of some form of "Pacific community," have been arrested by an unstated assumption of Canadian foreign policy that such actions would probably end up as a source of friction in Canadian-American relations and are to be avoided as far as possible. As a result, the trade aims and initiatives of Canadian governments in the Pacific have neither been buttressed by broader political relationships nor supported by a meaningful and comprehensive set of financial incentives. Ross's point is that Canada has sufficient existing sources of concern with the US and that Canadian governments simply do not wish to extend the range of contentious issues "unnecessarily." Yet acts of appeasement towards, or simply greater involvement with, the US have progressively reinforced Canada's vulnerability to US policies. It is not surprising, therefore, that the free trade negotiations with the US in order to deepen further economic relations between the two countries should be controversial.

Indeed, it may be argued that in the late 1980s proposals which suggest Canada should increase its trade with the US, that is for the Second Option Policy, are anachronistic. Given that for several decades Canada's trade has been overwhelmingly concentrated with the US and that tariff barriers between the two countries have already been comprehensively reduced, the benefits to further free trade may well be less than commonly supposed and probably much less than they would have been decades before (Watson, 1987). At the same time, the costs of adjustment to free trade may remain substantial (Watson, 1987). In addition, there are obvious difficulties facing trade enlargement with the US. One important difficulty relates to protectionist sentiments in the US, which have been spearheaded by an attack on the Canadian lumber (and shingle and shake) industry which has operated for the past 50 years within a free trade environment. The problem of continental trade enlargement is further underlined when it is realized that during the mid-1980s Canada enjoyed a $20 billion visible trade surplus with the US which the latter may wish to see reduced. That the US enjoys an equally large invisible trade surplus with Canada receives scant attention. In fact, invisible trade issues do not generate the same political implications and are less likely to be affected by changes in tariff structures than visible trade.

It may well be that some kind of negotiations are necessary simply to *preserve* Canada's relations with the US — the objective of the First Option. Canada could certainly benefit from a long term, structured approach to its economic relations with the US, although such an approach could probably be more effectively negotiated on a sectoral basis and possibly in the manner of the Auto Pact, a fixed rather than free trade agreement (de Wilde, 1986). But by pre-occupying itself with the US, Canada's bargaining power is necessarily circumscribed while, as Wallace suggests in the previous chapter, its economy may become even less competitive in global terms as the US becomes increasingly protectionist. And, if trade preservation with the US is the major continental goal of the trade talks, in what direction does trade growth lie? In this context, the continuing rapid rate of industrialization around the Pacific region has potentially considerable implications for Canadian trade patterns.

15.2 CANADA'S PACIFIC TRADE SINCE 1971

Canada's trade with its non-US Pacific partners increased steadily until the 1980s. Thus, non-US Pacific economies accounted for 8.2 percent of Canada's exports in 1971, 11.2 percent in 1981 and 10.6 percent in 1987. As such they have become more important than the EEC or the Pacific Region of the US (Alaska, Hawaii, Washington, Oregon, and California). Import trends from the Pacific region to Canada have been similar (Table 15.1). Yet, given the spectacular rates of growth achieved by several Pacific economies during the 1970s, notably Japan and the Asian NICs, some shift in the relative strength of Canada's trade relations with these countries was inevitable. In fact, closer ties might have been anticipated. Canada's trade participation in the Pacific region, for example, is much less than might be expected on the basis of the size of its total trade relative to the total trade of other economies and is more geographically specialized than any other important country in the region. This point is simply ilustrated by mapping visible trade links between pairs of countries whose actual trade exceeds expected trade, with the latter calculated on the basis of origin-destination independence (Figure 15.1).

Canada's only strong export trade ties are with the US. As the second largest export market, Japan accounted for a mere six percent of Canada's exports in 1979 which is less than might have been expected on the basis of the size of the two economies (Hayter, 1983). Only Mexico exhibits a similar degree of dependence on one market (also the US) but even Mexico has been able to develop export links to Central America and to China which are larger than expected on the basis of size considerations alone. Similarly, Australia's exports, which are more than expected to the US, New Zealand, Japan, China, and Hong Kong in terms of size of respective trade levels, are considerably more geographically diversified than Canada's exports. Remarkably, the trade figures for 1987 suggest that during the 1980s Canada's trade orientation to the US has increased while the shift to Pacific economies has been arrested (Table 15.1)

With respect to the structure of Canada's trade with the Pacific region, and excluding the US, exports are dominated by the food, feed and beverages group, especially wheat, fish, barley and meat; the inedible crude materials group, especially coal, rapeseed and copper; and by the fabricated materials

TABLE 15.1

CANADA'S VISIBLE TRADE WITH PACIFIC BASIN ECONOMIES 1971, 1981 AND 1987

	Exports			Percent	Imports	
	1971	1981	1987	1971	1981	1987
United States	67.8	66.2	75.7	70.1	68.9	68.0
(Pacific States)	(9.4)	(8.1)				
Japan	4.8	5.5	5.6	5.1	5.1	6.4
Australia	1.0	1.0	0.5	0.8	0.6	0.4
China	1.2	1.2	1.1	0.1	0.3	0.6
S. Korea	0.1	0.5	0.9	0.1	0.8	1.5
ASEAN	0.5	0.7	0.7	0.4	0.8	1.4
Other Asian	0.4	0.7	0.7	1.3	2.0	2.5
Mexico	0.5	0.9	0.4	0.3	1.2	1.0
Other L.A.	0.7	0.7	0.7	0.6	0.6	0.3
OTHER	23.0	22.6	13.7	21.2	19.7	17.9
Total %	100.0	100.0	100.0	100.0	100.0	100.0
Total $m	17,424	81,203	125,087	15,608	78,876	116,239

Source: Statistics Canada: 75,003. 65,00

group, especially coal, rapeseed and copper; and by the fabricated materials group, especially pulp, newsprint, aluminium, chemicals and lumber. Manufactured end products are a relatively small and declining component of Canada's exports to the region (Table 15.2). Australia, it might be noted, is the chief non-US Pacific market for Canada's manufactured end products. In contrast, Canada's imports from the region are dominated by such manufactured goods as autos, apparel, telecommunications equipment and industrial equipment principally from Japan, South Korea and Hong Kong. In terms of yearly fluctuations, these imports have been far more stable than Canada's exports (Hayter, 1983). In addition, these trade links have distinct regional orientations within Canada, with exports predominantly generated within British Columbia and imports primarily destined for Ontario and Quebec.

TABLE 15.2

CANADA'S VISIBLE TRADE WITH PACIFIC BASIN ECONOMIES (EXCLUDING U.S.) BY COMMODITY GROUP, 1971, 1981 AND 1987

	Exports		Percent		Imports	
	1971	1981	1987	1971	1981	1987
Food, Feed & Beverages	26.2	23.1	20.1	17.3	11.2	14.4
Crude Materials Inedible	32.4	29.2	26.7	7.2	14.0	24.6
Fabricated Material	26.1	33.7	34.8	24.1	13.5	11.1
End Products	14.9	13.5	9.2	50.5	60.1	49.9
	1,599	9,168	15,487	1,372	8,844	14,659

Sources: Statistics Canada: 165,003, 65,006

For definition of Pacific Basin Economies see Table 15.1

Notwithstanding the increased pace and scale of its trade relations within the region over the last 15 years, Canada's role within the Pacific region remains peripheral. Indeed, Canadian exports around the Pacific are frequently characterized in terms of untapped potential. The main failure, with respect to visible trade, has been Canada's inability to develop sustained large-scale exports of secondary manufactures. Canada's manufacturing exports around the Pacific typically originate in relatively small firms, many of whom export only on a sporadic basis (Hayter, 1986). Even the "committed" exporters normally serve specialized market niches for various kinds of capital rather than consumer goods. Not surprisingly, Canada has been unable to become a major supplier in support of the processes of industrialization occurring throughout the region. For example, China spent massively on imported plant and equipment during its third and fourth waves of modernization; in 1978 alone contracts worth US$6.4 billion were signed (Ho and Heunemann, 1983, p. 15). Very few of these contracts, however, benefited Canada even though in political terms China and Canada have well-established friendly relations. Canada has not been able to cash in on opportunities even when much political good will exists. The failure is widespread throughout the Pacific. As Kelly and LeCraw note, "between 1973 and 1982, Canadian companies did not submit a single bid on more than seventeen hundred civil works contracts financed by the Asian Bank with a total value of 1.4 billion dollars " (Kelly and LeCraw, 1985, p. 37).

As it happens, the level of competition for resource based markets throughout the Pacific is increasing as the traditional supplies from Australia, Canada, and the US have been augmented by newly developed sources in ASEAN, the USSR and New Zealand, while China also constitutes a potentially huge supplier of several resource-related goods. In addition, for the major resource consumers, notably Japan, supply is a question of politics and security as well as economics. Admittedly, Japan's search for diversified resource supplies has led to major investments in Canada. At the same time Canadian producers do get squeezed in international rivalry, for example, by trade negotiations between the US and Japan to the extent the US increasingly demands that Japan imports resources from the US to pay for its huge imports of Japanese manufactured goods. Whether or not Canada can significantly enlarge its manufactured exports — including to the Pacific region — is debateable. The problems are deep seated. In fact, over time Canada's global trade role has become entrenched by underlying patterns of investment.

15.3 INVESTMENT CROSSCURRENTS

On the whole, direct investments by foreign firms in Canada and direct investments by Canadian firms in other countries mirror and reinforce Canada's trading relationships. Thus direct foreign investment (DFI) in Canada is dominated by American firms and by two motives: to secure resources for the donor economy with a minimum of value added and to capture a share of the Canadian market for end products. On occasion, foreign investment is designed to capture some distinctive technological expertise developed by some usually small Canadian firm. In general, foreign-controlled subsidiaries have no direct interest or mandate to export secondary manufactured products from Canada (Britton and Gilmour, 1978). Indirectly, foreign ownership has further limited such exports by facilitating imports of components from affiliated sources and by restricting opportunities for Canadian firms to expand domestically to a size sufficient to justify investments in research, development and marketing which are necessary for large-scale exporting from a high income base.

Direct foreign investment (DFI) in Canada from Pacific sources, notably Japan, has further reinforced Canada's role as a supplier of resources and primary manufactures. Thus Japanese investments initially, and until the early 1970s, emphasized securing resource supplies for the home market from sources within western Canada. More recently Japanese plans, and those of Hyundai of South Korea, for investment in the auto industry clearly represent defensive motivations and are based primarily on serving the Canadian market. They have resulted at least in part from the pressure of the Canadian government on foreign companies to invest in Canada in proportion to the level of exports to Canada.

It is true that Canada has spawned a number of large transnational companies, perhaps more than is commonly supposed (Niosi, 1985) and some, for example, Alcan, B.C. Packers, Inco, Falconbridge, MacMillan Bloedel, Atco and Bata Shoes, have invested around the Pacific rim. With notable exceptions, such as the electronics operations established by Northern Telecom in Malaysia, most Canadian transnationals operate in traditional industries. A large number of Canadian companies, including many that are small and medium sized, have invested in the US but these investments are rarely seen as a stepping stone to investment in more distant and riskier environments. Certainly participation by Canadian companies in the Pacific region is not great. In ASEAN and South Korea, Tomlinson and Hung (1983a) could find only 207 Canadian companies with business activities within the region as of August 31, 1982. Moreover, only 85 of these companies, or 40 percent of the total, actually had equity investment in the region. These companies, it might be noted, were involved in a very wide range of activities including banking, consulting, shipping, transportation, sales and manufacturing. Many of these operations are very small. Indeed according to Tomlinson and Hung (1983b, p. 19), as a group the most important income earners for Canada were the engineering consultants. In total and in 1977, for example, Asia and Australasia accounted for C$1039 million or just 7.8 percent of Canada's DFI (Tomlinson and Hung, 1983a, p. 8). At that time, the two most important single recipients were Australia and Indonesia, the latter largely the result of the opening of a copper mine by Inco.

While DFI can imply a loss of exports and of jobs, in practice, it also generates considerable benefits to donor economies. Indeed, it is recognized that although exports may be partially replaced, or changed in nature, they are not eliminated by DFI. DFI can actually improve export potentials in various ways, for example, by providing a "captive" outlet for component parts and for various kinds of headoffice services, and by allowing companies to specialize in different stages of the manufacturing process. In some cases, DFI is essential for continued exports from the firm's home country. Foreign-based subsidiaries also impart to parent companies a better understanding of foreign market conditions and opportunities, which in turn facilitates future exports. In some instances, firms have to invest in a foreign market or lose export access to it. In addition, DFI provides companies with important flexibility advantages in terms of responding to change.

DFI, of course, also provides a source of profits. In fact, in macro terms, the available evidence suggests that the net effects of DFI on the balance of payments of such countries as the US, Japan and the UK are positive. Each of these countries, for example, has a substantial invisible trade surplus with Canada largely because of DFI. Unfortunately for Canada, it is probable that the advantages of DFI to the donor economy are not as strong in resource-based industries as in industrial and consumer goods industries. Indeed, some DFIs made by Canadian resource companies have been problematical to Canadian interests. Inco's copper mining investment in Indonesia, for example, constituted a considerable financial drain on the parent company's Canadian operations, and on completion contributed substantially to an over capacity situation. More generally, resource-based DFI does not offer the same opportunities for the export of components or for increasing market knowledge. Moreover, it has been suggested that resource based businesses are more vulnerable to the "obsolescing bargain" which refers to a process where the host country gradually acquires the necessary production know-how so that over time the relative bargaining power between the host country and transnationals increasingly favours the former. Typically, Canada's resource based companies have opted for joint ventures with local companies as a mechanism for entry into foreign countries. More recently, Canada's resource-based companies have been consolidating and even pulling out of their overseas operations, including those around the Pacific.

Yet in an increasingly interdependent world, exports and investments are complexly interrelated. A significant barrier in the Pacific to Canadian export hopes is the increasing degree to which trade links are underpinned by the activities of multinational corporations that are based elsewhere, notably in the US and Japan. The internationalization of trade within these corporations serves to eliminate export opportunities for the single plant Canadian firm. Branch plants of Japanese and American firms around the Pacific are also likely to purchase their inputs from affiliated sources; they provide an ongoing source of local market information for the parent company while demands for capital goods may be transferred back to the "donor" economy. Such behaviour clearly reduces export opportunities for the small Canadian firm operating solely from a Canadian base. For countries with small domestic markets such as Canada, the relationships between exporting and investment pose particularly difficult problems.

15.4 CANADA'S TRADE AND FOREIGN INVESTMENT POLICIES: A FUNDAMENTAL PARADOX

In the context of trade, Canada has generally supported GATT initiatives for a more liberal trading system and, for example, Canada was a leading supporter of Japan's request to enter the GATT framework (Langdon, 1983). Canada has also participated in the progressive reduction in tariffs and has been less protectionist than is sometimes recognized — and certainly less so than Australia, as Britton notes in Chapter 18. Canada, for example, did not seek to limit the flood of Japanese auto imports until they had accounted for 28 percent of the Canadian market, a much higher level than the 10 percent level permitted by the US. In terms of exports, federal and provincial governments have essentially attempted to provide a policy environment which is conducive to increasing exports without providing the kind of direct operating subsidies other OECD members might consider discriminatory. While Canada's export objectives have never been precisely spelled out, federal and provincial policy-makers have continually reiterated pleas for increasing the exports of manufactured goods including, and sometimes especially , in non-US markets. For many observers, increasing the degree of manufactured exports is *the* fundamental trade challenge facing Canada.

Canada has also favoured an extremely liberal policy towards DFI, caveats to FIRA and the NEP aside. Indeed, regardless of whether Canada's industrial policies have emphasized questions of regional balance or national competiveness policy-makers have almost invariably urged for an open door policy to foreign investment. In fact, throughout Canada there has been in place for some time a substantial political support system at federal, provinicial and local levels which attempts to facilitate the entry of foreign investment by the provision of advice, information on local conditions and various forms of incentives. Even within External Affairs, the orientation as regards investment is towards attracting business to Canada rather than on helping Canadian business to establish abroad. Moreover, there has not been systematic discrimination against Canadian acquisitions by foreign firms.

Free trade and an open door policy towards foreign investment are philosophically consistent. As even Japanese economists (e.g. Hayashi, 1981) point out, however, in Canada's case in pragmatic terms DFI is diametrically opposed to the goal of increasing exports of manufactured goods. In this regard, there has been a collective failure by federal and provincial policy-makers to appreciate the interdependencies between exporting and DFI. Simply stated, DFI is seen only as a benefit to host countries. Host country costs are discounted. Similarly, the benefits of DFI to donor countries are ignored and government policy regarding DFI by Canadian firms is ambiguous (Litvak and Maule, 1981). There are certainly no policies which directly support such investment, except in the very limited instances of Canadian International Development Agency (CIDA) suppported industrial co-operation agreements which are part of Canada's development effort. Government policy statements regarding DFI by Canadian firms are noteworthy only by their absence, and perhaps such behaviour is another example of a "prudential evasion of formal commitments." Possibly, DFI by Canadian firms is seen by policy-makers, at best, as a mixed blessing or even entirely in negative terms from political, social and economic points of view. If so, there is a considerable underestimation of the benefits of DFI by Canadian governments to the Canadian economy.

Whether or not Canadian policy-makers can resolve the need for short-term jobs, perceived to be strongly dependent on foreign investment, and the need for long-term jobs, perceived to be dependent on increased manufactured exports, is problematical. For almost two decades the Science Council of Canada has offered numerous suggestions to stimulate a more innovative and export-oriented manufacturing sector (e.g. Britton and Gilmour, 1978). By and large, however, these suggestions have not been implemented with real conviction, if at all (see Britton, Chapter 18). Similarly, the idea of a National Trading Corporation to help promote the export potentials of small firms has been mooted from time to time, but in policy terms it remains an ambiguous concept (MacMillan, 1981; Kelly and LeCraw, 1985; Klein, 1987).

To the extent that Canada has permanently accepted its status as a branch plant economy, Canadian initiatives in the Pacific will remain limited and its role restricted to that of resource supplier. In this regard, Canada's striving for (more) free trade with the US further confines, as Terry McGee suggests in Chapter two, Canadian policy to simply hanging onto the coattails of the US.

Changes in attitude and of policy are nevertheless possible. It may be, for example, that the continental free trade/coattail policy will come to be seen as the policy of the past offering nothing new in the form of ideas and ambition for Canada's future global role. One possible cause for such a change in perception relates to the colossal balance of payments deficit facing the US, which is increasingly seen as the crux of world economic problems and which may well create difficulties for Canada in two immediate ways. First, and most obviously, pressure may be brought to bear on Canada to reduce its visible trade surplus with the US. Second, if, to resolve its trade problems the US has to increasingly share its global economic leadership with Japan and the EEC, then Canada's position may become even more vulnerable and Canadian ideas of a special relationship with the US simply another illusion. If Canadian ambitions should broaden, for the foreseeable future, the dynamism of the Pacific economies will continue to provide Canada with alternative trade and investment strategies.

16. IMPLICATIONS OF PACIFIC INDUSTRIALIZATION FOR AUSTRALIAN STRUCTURAL CHANGE

Peter D. Wilde

Looking back from the start of the 1990s, it appears that the 1970s was a decade when global transformation and its national repercussions took Australians by surprise, while the 1980s was a period when the nation tried to come to terms with change only to find as the decade advanced that the rules themselves were changing and that continued transformation was needed to keep abreast of new realities. Nowhere was this changing perspective more apparent than in the nation's international trading and diplomatic affairs, which saw an accelerating shift of focus from Britain and other European countries to Australia's more immediate neighbours in Eastern Asia and the Pacific. By the 1980s that shift was virtually complete, but the prize promised by that integration into the rapidly growing Pacific economy remained as elusive as ever, as the dynamics of the global economy and body politic continually restructured. The 1980s also saw a maturation of attitude by many in Australia, and a growing realization that the complex interweaving of global, national and local processes makes national and state governments relatively powerless to control events within their territories, even though — for good or ill — their influence can be critical.

The purpose of this chapter is to examine Australia's changing trade relations at a global level, particularly in relation to the growing economies of the west Pacific. The chapter also explores the hopes for the Australian economy regarding integration with the Asia-Pacific region and the fears that are held about the capacity of various groups within Australia to prevent that integration from occurring.

As Linge indicates in Chapter three, the Australian economy has recently reached a crisis point. By the early 1970s it was clear to many commentators that the structures that had served Australia well since before the turn of the century were no longer applicable. Major planned and unplanned structural change has occurred in Australia since 1970, in part because of the major restructuring of the global economy itself (Fagan, 1987; Linge, 1988; Rich, 1987, 1988). Central to these now outdated structures were the export of primary goods, isolation from the rest of the world of the manufacturing and service sectors, and wage fixing and financial systems comprising a grotesque, *ad hoc* and ever growing array of rules, restrictions and regulations. Nevertheless, short-term recoveries and mineral booms continued to weaken enthusiasm for change, and successive governments, apart from rarely implemented desires to reduce the effective protection given by tariffs, took few real steps to encourage structural change. Most manufacturing firms, with attitudes developed in the regime of high tariff protection, continued to operate in the traditional manner, producing for the home market and treating exports as a marginal addition. Given that many companies are offshoots of transnational corporations — deliberately established behind Australia's tariff barriers to gain a share of that market, and sometimes explicitly forbidden to export, and given the fragmentation and inefficiency of manufacturing generated by the tariff policy and the internal political and economic structures (Linge, 1979; Linge, Chapter 3 this volume), this failure to change direction is not surprising. Organized labour, for its part, continued to use its considerable power to increase its share of GNP whenever it could.

By the mid-1980s, it seemed at least possible that the inadequacy of short term booms to solve the deep-seated problems of Australia, and hence the urgent need to initiate long-term change, was sufficiently recognized by political, business and union leaders, and perhaps even by the larger community, for some commitment to action to be given. The latter half of the 1980s has, indeed, seen some halting, some imaginative, some brave and some foolhardy steps by government, business and unions to restructure the Australian economy to meet the changed and changing global situation.

16.1 PACIFIC INDUSTRIALIZATION AND TRADE

Economic growth of some East Asian nations for the two decades prior to the early 1980s was outstanding. In 1962 they collectively accounted for 10 percent of world GNP. Twenty years later they had increased this share to 15 percent, while North America and western Europe had each lost two or three percentage points to around 27 percent each.

Economic growth in several East Asian nations was largely brought about by shifts of policy from import substitution to export orientation. In some cases exports were predominantly of natural resources while in other cases the Japanese model of manufacturing for export from imported materials was followed. While the details of policy, association with transnational corporations, investment incentives and treatment of labour vary among the countries, the outcomes have been largely similar and, in the initial stages at least, the newly industrializing countries were helped by the preference system for less developed countries allowed by GATT.

As a consequence of these policies, exports from East Asian countries almost doubled to 17 percent of GNP in the 20 years to 1982, with such countries as Korea and Taiwan, showing particularly large increases. Over the same period the region's share of world exports and imports both more than doubled to 18 percent, accounting for about half of the Pacific total. Thus, in the early 1980s the Asia-Pacific was the world's fastest growing region in terms of output and trade, and some commentators suggested that the centre of world economic power was rapidly moving from the Atlantic to the Pacific.

This rapid growth of a number of national economies, each dependent on exports to the rest of the world, was inevitably complex. The economic growth of the region as a whole, and to a large extent of individual countries, depended on a measure of complementarity and non-competitive specialization in their economies. But with industrialization came constant and rapid structural change, the dynamics of which were felt throughout the region. Real wage levels and skills were seen as a major driving force in this restructuring, for as these rose, countries were forced to shift their export sectors to more skill and capital-intensive products in order to remain competitive against both current producers and against lower wage countries. Low-wage countries, in the region and beyond, were then able to promote their own development by producing and exporting labour-intensive products. Japan began this transition in the 1960s and Singapore, Hong Kong and Taiwan were able to expand their labour-intensive exports, notably of textiles and clothing. As these countries in turn gradually built up more skilled and higher paid work forces and more capital-intensive economies, then some of the ASEAN countries and China — until the dramatic upheavals of 1989 — attempted to increase national income by using their generally low-skilled but very cheap labour forces to produce light manufactured goods for export. Japan is, of course, still leading these countries in the sophistication of its technology and products; the emphasis in Japan on a shift to high technology and skill-intensive goods is matched by a reduction in the importance of some capital-intensive industries. For example, the steel industry, although technically advanced by comparison with that in many other advanced countries, is already past its peak and steel is unlikely to be a significant product by the year 2000. Taiwan, Korea and even China, which in the early 1980s had small steel industries, began building up capacity with the aim of becoming major producers in Japan's place (Anderson et al.,1985; Edwards, 1982).

As an outcome of these changes, the share of labour-intensive goods in the exports from the west Pacific as a whole increased from 20 percent in 1960 to 25 percent in 1985. Over the same period, Japan's exports of labour-intensive products declined from 13 percent to 5 percent and its imports of these goods began to increase. By the early 1980s Hong Kong, Singapore, Taiwan and Korea had more than filled the gap left by Japan's changing industrial structure but were already well advanced with the diversification of their own economies towards more sophisticated and capital-intensive export products. As the decade advanced, China and some of the ASEAN countries assumed the role of cheap labour havens and exporters of labour-intensive goods.

It would be a mistake, however, to think that all countries follow similar dynamics of development on the same ladder of technological progress. Internal political and institutional situations vary considerably, as do policies of industrialization. Moreover, politicians and businesses in the countries which have abundant natural resources, notably oil, such as Indonesia, Malaysia and Brunei, are less likely to be motivated to pursue industrialization with as much determination as those in countries without such assets. In addition, the process of structural change and the opportunity for less developed countries to share in industrialization are in danger of being hampered by tariff protection and restrictive bilateral agreements in more developed countries. For example, armed with the threat of even tougher legislative restrictions, the US concluded "voluntary" textile agreements with each of Korea, Hong Kong and Taiwan restricting the growth of their textile exports to US to one percent or less per year. Australia has also protected a number of industries in order to permit so-called orderly restructuring.

By the middle of the 1980s, however, it was clear that the rapid economic growth in those newly industrializing Asian countries which had so successfully used the restructuring of global manufacturing to their advantage was beginning to flag and the region's future was at best uneven growth and at worst economic decline. By the end of the 1980s it was apparent that the reality is that these countries are caught, by their geographical location and their trade patterns, in the middle of the Japanese-US trade and financial battles, and there is little that they can do to influence the outcomes. The Asia-Pacific region will remain important and active in the global economy, but it will not become the shaper of the world's economic future as so many prophesied in the 1970s (Daly and Logan, 1989, p. 252).

16.2 CHANGING PATTERNS OF AUSTRALIAN TRADE

Australia has long had an export trade largely dependent on rural and mineral primary products and a heavily protected, inefficient and inward-looking manufacturing sector which, since the early 1970s has faced a severe reduction in the face of growing imports (Wilde, 1981). During the post war period there has been an increasing shift of trade relations away from Europe and towards Northeast and Southeast Asia. In the early 1950s, scarcely 15 percent of Australia's exports went to these regions and less than 10 percent of imports came from them. By 1965 almost 30 percent , and by the end of the 1980s virtually half of Australian exports went to Northeast and Southeast Asia, while imports from these regions increased from around 16 percent in 1965 to almost 40 percent by 1989 (Table 16.1). In 1988, 45 percent of Australia's trade was with its four major Asian partners and by the mid-1990s it is expected that 60 percent of Australia's export earnings will come from Asia (Daly and Logan, 1989).

Most of these changes were at the expense of trade with the United Kingdom, which was declining even before admission to the European Community (EC) dramatically increased trade barriers and other restrictions. From an average of 36 percent in the early 1950s, the United Kingdom's share of Australia's trade fell to 11 percent around 1970 and to under 5 percent by 1980 where it remained throughout the decade. At its simplest this redirection of trade reflects the growth of the East Asian economy and the complementarities that Australian primary production has with the manufacturing sectors of some of the countries of the region.

TABLE 16.1
AUSTRALIAN IMPORTS AND EXPORTS BY COUNTRY AND REGION, 1965 - 1989

(a) Imports				Percent			
	1965	1970	1975	1980	1985	1987	1989 (Jan-July)
China	0.8	0.8	0.9	1.2	1.2	1.9	2.2
Hong Kong	0.9	1.5	2.3	2.2	2.1	2.1	1.7
Japan	9.6	12.9	17.9	17.0	23.1	19.3	21.1
Republic of Korea	0.1	0.1	01.6	1.0	1.5	2.5	2.7
Taiwan	0.2	5	1.4	2.7	3.4	4.3	3.9
(Northeast Asia	11.6	15.7	23.1	24.1	31.3	30.1	31.6)
ASEAN	4.1	2.6	3.3	6.9	4.9	5.7	6.3
New Zealand	1.5	2.4	2.7	3.4	4.1	4.0	4.1
North America	27.3	29.4	22.3	24.5	23.7	23.2	23.6
Western Europe	42.7	40.3	35.8	26.6	26.6	28.1	26.8
Rest of World	12.8	9.8	12.9	14.7	9.5	9.0	7.6
TOTAL	100.0	100.0	100.0	100.0	100.0	100.0	100.0

(b) Exports				Percent			
	1965	1970	1975	1980	1985	1987	1989 (Jan-July)
China	5.5	2.9	2.8	3.7	3.9	4.0	2.9
Hong Kong	2.1	1.5	1.1	1.4	1.5	2.1	3.8
Japan	16.6	27.7	30.5	25.2	26.0	24.0	26.3
Republic of Korea	0.2	0.3	1.5	2.2	3.9	4.1	5.3
Taiwan	0.6	0.9	1.11	1.9	3.0	33.6	3.5
(Northeast Asia	25.0	33.3	36.9	34.4	38.3	37.7	41.8)
ASEAN	4.3	6.0	7.7	7.7	6.4	6.0	9.3
New Zealand	6.3	5.1	5.0	4.5	3.6	4.6	4.8
North America	12.4	15.8	12.5	11.9	8.8	10.2	11.6
Western Europe	35.0	23.7	17.2	14.1	13.5	16.4	15.5
Rest of World	17.0	16.2	20.7	27.5	29.5	25.1	17.0
TOTAL	100.0	100.0	100.0	100.0	100.0	100.0	100.0

Note: Totals may not add due to rounding
Source: Garnaut (1989)

Of the many nations of the Asia-Pacific, Japan is overwhelmingly important for Australia's trade, although the US ranks a fairly close second as a source for Australian imports. Next in importance are those few newly industrializing countries (NICs) of the west Pacific: Hong Kong, Singapore, Korea, and Taiwan and, considerably further behind, the ASEAN nations of Brunei, Indonesia, the Philippines, Malaysia and Thailand, which are generally also undergoing industrialization, though are at an earlier stage than the northwest Pacific NICs. China grew steadily in its importance as a trading partner until the mid-1980s but given its enigmatic politics, its role in the 1990s is unpredictable. Australia's trade with New Zealand, in value terms scarcely below that with ASEAN as a whole, and with the former Australian administered protectorate of Papua New Guinea, has quite different characteristics and significance. New Zealand is the destination of a high proportion of Australia's exports of manufactured goods, machinery and equipment, while food products and crude materials rank high among imports . This specialization between two neighbouring countries which share very similar histories and positions in the world economy is likely to increase as the "Closer Economic Relations" agreement is progressively acted upon (Australia, Senate Standing Committee on Industry and Trade, 1985). Most of the trade with Papua New Guinea, on the other hand, is related to aid agreements, and though these are still relatively numerous, Australia is actively reducing them. The remaining nations of Asia and the Pacific have very limited trade relations with Australia.

Despite this shift in trade towards the world's fastest growing region, Australia has been unable to maintain its trading position. The share of exports in Australia's GNP declined from 14 percent in 1962 to 13 percent in 1981 at a time when the Pacific as a whole increased its share from 5 to 28 percent and the world from 9 to 16 percent. Australia's share of world merchandise trade consequently declined from around 1.9 in 1965 to less than 1.2 in 1987, despite the fact that during the same period the Pacific as a whole increased its share from just over one quarter to almost two fifths, within which North America remained roughly constant. In terms of rank, Australia slipped from being the world's twelfth largest trader in 1973 to its twenty-third by 1983, and its shares in the import trade of these growing Asia-Pacific nations has fallen since 1965.

The commodity structure of Australia's exports has always been heavily biased towards the primary sector, which since 1970 has accounted for just over 60 percent of export value. Between 1970 and the mid-1980s, the share of manufactured goods (a category which includes non-ferrous metals) declined from about 20 percent to under 17 percent of exports, while services and other exports overtook manufacturing by growing from about 17 percent to 20 percent (Rimmer, 1986). This structure was very different from the trade pattern of the world as a whole where manufactured goods accounted for around 50 percent, primary products have fluctuated around 30 percent and services and other goods for slightly less than 20 percent (Table 16.2) .

TABLE 16.2

AUSTRALIAN AND WORLD EXPORTS BY SECTOR 1972 - 1984 (A) (B)

	Australia (percent)			World (percent)		
	1972	1980	1984	1972	1980	1984
Agriculture	45	39	31	16	12	12
Fuel	5	9	19	8	19	16
Other minerals	13	16	13	2	2	2
Total of above	**63**	**64**	**63**	**26**	**33**	**30**
Manufactures (c)	20	19	17	53	48	52
Services	14	14	14	19	17	17
Other	3	3	6	2	2	1
Total	**100**	**100**	**100**	**100**	**100**	**100**

(a) Estimated for 1984
(b) Classification based on SITC revision 1
(c) Includes non-ferrous metals and alumina for the world but not Australia
Source: Australia, Economic Planning Advisory Council (1986)

There has been a marked change, however, in the composition of Australia's primary exports. Those to Britain were almost entirely agricultural products, with wheat and wool dominating. As the importance of trade with East Asia increased, agricultural products remained significant, but coal, iron ore and other mineral products grew to be of roughly equal importance, and it is clear from Table 16.2 that most of the growth, especially in the early 1980s, was in fuels. Indeed, while Australian trade appears highly specialized at this broad aggregate scale, and often draws unfavourable comparisons with Argentina,

Mexico or Malaysia (Duncan and Fogerty, 1984), its primary sector exports are in fact highly diversified. On the other hand, the volume of demand and the world prices for minerals are notoriously volatile so that both export prices and capital works programs have a violently cyclical pattern. While variations in price or demand for individual commodities may have little effect on the Australian economy as a whole, a general oscilation, such as that experienced in the 1980s, is bound to have repercussions throughout the economy.

The export trade with East Asia began in the late 1950s essentially because Japan's growth allowed the development of Australian mineral, and especially coal and iron ore, resources. Japan was locked out of most of the world's mineral resources by the ownership and strategies of vertically integrated mining and processing companies and the difficulties of obtaining new resources in less developed countries. Australian mineral resources, however, had been left undeveloped by the transnational corporations which owned them since the European and American users had alternative supplies and the costs of transport to traditional markets were excessive. Since Japan at that time was wholly committed to developing mineral processing, had few skills in the mining field and imposed restrictions on capital exports, long-term contracts were drawn up with Australian developers as the only way to ensure the security of demand which would warrant the development of the resources. Neither side had much flexibility once the contracts were drawn up and operations were so highly vertically integrated and interdependent that the contracts explicitly stated how risks were to be shared and surpluses distributed among the the participants. Frequent renegotiation of detail has been essential as conditions change, but the basic agreements have remained.

Although Japan was initially extremely dependent on Australian cooperation, the relationship has generally been regarded as more important and inviolable by Australians than by the Japanese. On occasion this has led companies, and more usually the labour unions, to try to exert pressure on the Japanese to increase prices. As a pro-active strategy the Japanese encouraged mining developments elsewhere. For example, British Columbia was assisted in developing coal mines in order to diversify supply and create a bargaining point, despite the fact that, at exchange rates prevailing at the time, delivered prices would be 50 percent higher than Australia's. Resources in less developed countries around the Pacific are also being developed as part of the same strategy. In addition, increasing direct foreign investment by Japan, encouraged by the highly valued yen, gives rise to the possibility of establishing foreign branch plants to process the minerals. The broad outcome is that, while Australian exports to Japan have increased absolutely over the years, they have not maintained their share of Japan's imports. Despite quite violent fluctuations from year to year, however, they have, since 1969, retained an average share of Australia's exports of rather over one quarter. In the region as a whole market conditions are constantly changing, and Australia's share of regional trade may well change more markedly than hitherto.

Like export destinations, Australian sources of imports showed a similar shift towards the Asia-Pacific region (Table 16.1). In 1965 around 55 percent of Australia's total imports came from the region and this increased to about 65 percent in the last half of the the 1980s. Japan alone increased its share from less than 10 percent in 1965 to over 20 percent in 1988. America's share of Australian imports declined from almost 30 percent in the late 1960s to around 24 percent two decades later, but the most significant decline was for the UK; its share declined from over 20 percent in the late 1960s, itself a major decline from the 45 percent of the early 1950s, to around 7.5 percent in the last half of the 1980s.

As a result of the large volumes of primary products exported to east Asia, Australia usually maintained a bilateral trade surplus with most individual countries, including Japan, with whom, for example, there was in 1984-85 a A$1,200 million surplus, compared with the US deficit of some US$40,000 million. These trade surpluses did not stand well alongside Australia's high tariffs and quota system on the very products which the countries of East Asia, especially those industrializing most recently, wished to sell on the Australian market. Inevitably there has been constant tension between Australia and its trading partners who frequently attempt bilaterally or collectively to coerce or bargain with Australia to obtain easier access to the market. Successive governments have found it difficult to balance the political and economic claims of Australia's trading partners and the owners and employees of the sensitive domestic industries which are put at risk by the reduction of protection. While protection has been reduced, and many manufacturing jobs lost as a result, particularly in the textile and clothing industries, protection nevertheless remains high. Continuing changes in tariff and other forms of protection, along with technological and organizational restructuring will almost inevitably give rise to further periods of dramatic change in work practices, occupational structure and levels of employment in manufacturing throughout the 1990s.

Important though trade in goods and services remains in Australia's external relations, the 1980s closed with more attention focussed on international financial matters, with prominence given to exchange and interest rates, capital flows and indebtedness. The 1980s saw a massive rise in national indebtedness from an annual average of about A$650 million in the early 1970s to almost A$22,000 million in 1986-87, sufficiently large to place Australia among the top debtor nations of the world. Associated with this increase was a decline in the share of the capital inflow made up of direct investment from 80 percent in 1975 to 20 percent in 1986, and an increase in portfolio investment and large loans raised by private sector foreign- and Australian-owned enterprises alike (Fagan, 1988). By 1986-87 the importance of mining and manufacturing as recipients of foreign capital had declined, while over half of the capital inflow was for finance, property and business services (Crough, 1988). Yet, despite the record levels of foreign investment, new investment in productive plant and equipment fell steadily from 1975 (Lougheed, 1988). The flights and fancies of global capital have become at least as important a consideration as production and commodity trade in Australia's economic management and national well-being.

16.3 AUSTRALIAN TRADE POLICIES

By the mid-1980s there was at last a fairly broad understanding in Australia that substantial structural change was occurring in both trading relations and the domestic economy. Most importantly, the nature of the interdependencies between the exports of primary goods on the one hand and domestic consumption and imports on the other were beginning to be appreciated. A commitment to closer integration with the Asia-Pacific region became more widely held, not least because of the persistence of the Hawke Labor government from 1983. The need for a more positive attitude to exporting and a deliberate policy and direction for industrial restructuring within Australia became more widely accepted.

A crucial debate focussed on whether a high level of primary exports should continue to form the core of the nation's trade policy or whether there should be a drastic restructuring to encourage manufacturing and services to fill a greater share of exports. Undoubtedly primary products will continue to be the mainstay of Australia's trade for the foreseeable future and, even with a committed development of the other sectors, it will be many years before they approach the importance for Australia that they have in world trade as a whole. Nevertheless, the two approaches represent rather different philosophies.

Those who propose concentrating on primary products support their argument by pointing to Australia's current comparative advantages and trading patterns. They argue that this emphasis represents an efficient allocation of resources, which would be even more evident if the inefficiencies inherent in the protection of the manufacturing sector were allowed to wither away. Australia's future prosperity, they argue, depends on competitively delivering these products to its Pacific (and other) trading partners, and the nation's standard of living will increase hand in hand with their partners' trading success with the rest of the world.

This approach has two major problems. The first is that in the long term the terms of trade for primary products have worsened. It is well established that the proportion of income spent on foodstuffs (especially basic foodstuffs like most of Australia's) declines as incomes rise. Further, the demand for mineral products also declines over time for any given level of world economic growth, as changing technology, product usage, high technology substitutes, increasingly sophisticated products and so on reduce the mineral and material content of final goods. Thus there has been an average one percent per year reduction in the metal content of products since the beginning of the century and this rate of decline is probably accelerating. It is undoubtedly a long-term phenomenon. As a result, Australia's terms of trade, while fluctuating in the short-term, are persistently worsening in the longer term and by the late 1980s were about one third lower than in the mid-1950s. To rely on the export of primary products to increase living standards thus requires an ever increasing share of world trade in these products in order to offset the reducing value of imports that can be bought for a given level of exports (Australia, Economic Planning and Advisory Council, 1986).

The second problem concerns the general trading regime for primary goods and metals in particular. World prices for primary products are notoriously volatile in the short term making it extremely difficult to maintain a consistent long term policy when immediate events have dire consequences for the balance of payments, overseas borrowings and debts, the exchange rate, inflation and monetary, fiscal and employment policy. Inevitably, short-term economic management took precedence over, and in part destroyed, long-term policy initiatives during the 1980s and will doubtless do so to an even greater extent in the 1990s as the Australian economy becomes even more open to external influences.

Further, the behaviour of Australia's competitors and customers exacerbates these problems. Many mineral exporters among the less developed countries are frequently insensitive to prices or profits, since the high foreign debt they must service leads them to increase output with lower prices, often driving the price down further. There are also many large, very competitive mines, highly capitalized with modern technology which will come on stream if mineral prices improve and further cut into Australia's markets, profits and growth. Thus the role of mineral production in Australia's Gross Domestic Product could decline in the event of substantial price shifts in either direction.

Among Australia's customers, restrictive practices and discriminatory controls limit markets. Agriculture has few agreements within GATT that encourage free trade, and protection is high and variable in Japan, the US and the EC. Australian exports are inevitably reduced, and retaliatory actions between the US and the EC have been shown to have adverse effects on Australia's sales despite the unwillingness of US governments to acknowledge the evidence. Tariffs designed to maintain home processing of minerals also abound; for example, the EC's differential between tariffs on bauxite and alumina. When these restraints are added to the usual strategy of mineral processors to retain flexibility of supply by committing as little capital and making as few long-term agreements as possible to raw material sites, then the possibilities of increasing Australian GNP from downstream processing are drastically reduced.

The alternative policy approach of developing the manufacturing and service sectors, possibly at the expense of the primary sector, is put forward by those who believe that these sectors are the only ones which offer hope in the long-term for Australia to maintain its rank in world living standards. They argue that the growth in value of exports of rural and mining products will not counterbalance the long term adverse movement in the terms of trade. Only in the manufacturing and service sectors is job growth considered likely. The rural and mining sectors, in fact, employ only 8 percent of the work force, are strongly biased towards males, and their capital intensity is constantly increasing.

Those who argue for increased exports of more sophisticated, elaborately transformed manufactured goods and producer services recognize that these sectors too are becoming increasingly capital intensive, and that Australia's capacity to export depends on adopting competitive technology. World trade in these sectors is increasing rapidly (for example, see Chapter 5 by Langdale), and the terms of trade are moving in their favour. Australia has only a small share of its exports in these sectors and so, it is argued, as a small producer in world terms, it should be possible to find a series of niches in the market which do not threaten other producers but which have a big impact on the Australian economy.

All these matters are ripe for political expression and exploitation. The issue of whether federal government funding and policy should favour the rural sector and mining companies or should concentrate on the large proportion of the population that lives in urban areas is a particularly divisive issue between and within political parties. The tussles within the parliamentary Liberal Party between Peacock on the one hand and Fraser and Howard on the other epitomize the ebb and flow, and perhaps most recently the muddying, of this issue. Funding policies have considerable distributional implications between places and social groups, especially resulting from government infrastructure spending, foreign investment regulations, and manufacturing policy. The left wing of the Labor Party and the wetter sections of the conservative parties tend to have urban electoral bases, and are inclined to favour orderly restructuring and the relatively slow dismantling of protection in order to protect urban jobs. They also seek the maintenance of a high level of government funding in urban areas generally. The right wings of all political parties are more favourably disposed towards the inflow of international capital and call for further deregulation to give closer integration with the world economy, albeit as a babe amongst giants. The latter half of the 1980s saw a steady shift to the right in both the federal Labor government and the conservative opposition, a shift confused and to some extent tempered by increasingly vociferous, organized and politically potent environmental groups. These divisions within and between the major political parties and the attempts by both traditional and new parties to woo the environmental and New Age vote make long term planning and community acceptance of and confidence in any strategy very hard to achieve.

Not all those who wish to see the growth of the manufacturing sector's employment and output favour an export-oriented policy. Some wish to maintain the *status quo* in terms of continued tariff protection and industrial and regional economic structure, often from fear for existing jobs and power bases. Some, mostly from the academic left, strongly express the view that, as a relatively powerless nation, Australia should remain economically and politically as independent as possible from the rest of the world and the power of other nations and transnational corporations (Crough and Wheelright, 1982).

The present federal government has been firmly committed to a policy of developing the export capacity of manufacturing and services, particularly with the Asia-Pacific market in view. Even so, community hardship, political pressures, changing corporate strategies, the demands of short term economic management and the growing importance of environmental responsibility make it very difficult to pursue a consistent or credible long-term policy. The steel and car plans, for example, were designed to restructure and improve the competitiveness of, respectively, a monopoly and an over-diversified industry (Wilde,1986). Nevertheless, increased protection, intended to be for the short-term only, was granted, though the plan for the car industry was soon overtaken by the global strategies of corporate restructuring. Early in the government's term of office the then minister for Science and Technology enthusiastically launched a national technology strategy (Australia, Department of Science and Technology, 1984), which particularly promoted increased advanced education and training. A review of Australian science and technology by the Organisation for Economic Co-operation and Development (Organization for Economic Co-operation and Development, 1985) recommended the use of new technology in high value added sections of Australian agriculture, fisheries and chemical industries. In fact, these proposals have had few practical outcomes (Carter, 1986).

During the 1980s, however, significant and far-reaching changes were also being made in the service and financial sectors. Some firms engaged in tradable services, notably construction and consultancy developed major operations offshore. Considerable deregulation in banking and other financial services occurred and a limited number of foreign banks were given licences to operate in Australia. The Australia dollar was floated in December 1983, soon after the Hawke Labor government assumed office, and the decision strongly defended during its volatile and often speculative movement over the years since the float. In the six years following the float, the Australian dollar traded as high as 95 US cents and down almost to 60, while at the end of 1987 it would buy only 95 Japanese yen, well under half its purchasing power in 1983 and 1984.

Despite a continuing commitment by the Australian government to restructuring the domestic economy and integrating it more closely with the Asia-Pacific region, considerations of political, social and short-term economic realities make it very difficult to pursue appropriate policies in a consistent, long-term manner. This makes it even more difficult to change entrenched attitudes in many parts of the business community and unions. The long history of protection and the current erratic policy environment ensure that many Australian companies have not seriously considered expanding beyond the domestic market. They are unenthusiastic about exporting, and few executives expect their firms to grow at the same rate as their Pacific market (Committee for the Economic Development of Australia, 1986). Much to the dismay of federal and state governments, their approach is fragmented with little interfirm cooperation or concentration on particular product or territorial markets. Individual firms may have overseas sales agents, but few have established their own sales office, and sales are spread thinly over countries where Australian products are marginal competitors rather than market leaders. There is, for example, a general lack of Australian firms, just as Hayter noted with respect to Canadian firms in the previous chapter, in the major developments taking place in Bangkok. A report on non-price competitiveness had Australia in mid-rank among advanced nations on product safety, marketing, after sales service and of lower rank on quality and a willingness to modify products to suit the local market. A few Australian firms find market niches and become market leaders for particular products or services in particular countries but most remain primarily concerned with profit viewed in terms of cost competitiveness and hence avoid the uncertainties of foreign exchange transactions. To change the habits — entrenched over decades — of an entire manufacturing sector, is an enormous task, especially when the international ground rules are constantly changing.

On the other hand place and nation-space is irrelevant to capital which is global in its perspective, and mergers and takeovers occur regardless of the national origin of firms or the financial status of nations from which bids originate. Australia, an indebted and capital-absorbing nation, has been the seat of a number of international takeover bids by corporate entities such as Broken Hill Proprietry Ltd. (Fagan, 1984) and Elders-IXL Fagan, (1990) and some one-time Australian nationals such as Alan Bond, Rupert Murdoch and Robert Holmes-a-Court have achieved a reputation for playing the global capital market with greater or lesser panache and success.

16.4 CONCLUSION

Unless Australia's domestic economy and its relationships with the dynamic Pacific region are dramatically restructured, the nation faces a dual problem. On the one hand, as Treasurer Keating honestly but perhaps unwisely warned in 1985, Australia is likely to suffer a reduction of living standards which in time will give it a status little better than a third world country. On the other hand, unless the urban majority shares in the wealth of the country, most probably and simply in the current situation by having access to jobs and higher living standards, then the social and political cohesion of Australia is at risk.

This message has been given with increasing intensity and urgency over the 1980s, and there are signs that it is at last being heeded. The attitude of the Labor federal government, its "Accord" with the union movement and the acceptance of its labour and industry policies by a broad section of business people have given an opportunity, rare in Austral's short history, for capital, labour and the state to co-operate on a national scale to face the competition on an international scale (see, for example, *Business Review Weekly*, 1989). Yet that co-operation is fragile and constantly threatened from the left, centre or right of politics, and the national election in early 1990 may lead to quite different policies, reflecting a different balance of power and personality within and between political parties.

Consistent, long-term and widely-accepted policies, while dreamed of by many, are not the stuff of the Australian state, nor if the 1980s are any guide are they the key to national success in an era of global change. Australians have learned slowly and painfully that the political and economic realities of their global geography must take precedence over their historical and cultural ties, that events in the Asia-Pacific region are more important than those in a Britain about to enter a unified European Community. It would be ironic indeed if the salvation so fervently and recently promised to extend from that change of relationships proved to be an illusion.

ACKNOWLEDGMENTS

The author is grateful for comments from Godfrey Linge on an earlier draft of this paper.

17. WHITHER REGIONAL POLICY? AUSTRALIAN AND CANADIAN PERSPECTIVES

Peter McLoughlin and James B. Cannon

Regional policy is a relative newcomer to the public policy arena. It gained prominence during the long economic boom in the post-World War II period. As a new claimant for public sector resources, regional policy met resistance from older entrenched policy establishments. However, during this period of public sector expansion, the political commitment to regional policy was strong enough for it to win grudging acceptance. Nevertheless, the relationship between regional policy and older more established policy fields remained uneasy. In particular, the territorial orientation of regional policy often resulted in conflicts with other policy fields which tended to have more traditional functional bases (Friedmann and Weaver, 1979; Lithwick, 1982b; McLoughlin, 1986a). Moreover, events of recent years suggest that the flowering of regional policy may have been short- lived and that it is now in retreat in both Australia and Canada.

This chapter first examines current directions and future prospects for regional policy in Australia and Canada. In the subsequent section, the increasingly tenuous position of regional policy on Australian and Canadian public policy agendas is documented and reasons for this development are discussed. This analysis provides a context for considering future policy directions in the concluding section.

17.1 DECLINING STATUS OF REGIONAL POLICY

The retreat from regional policy is apparent in a variety of ways in both countries. In Australia, the present status of Commonwealth (federal) regional policy is piecemeal and displays all the hallmarks of ad hoc incrementalism and political expediency. Explicit Commonwealth regional policies fall into two categories. First, there are those which owe their existence to an earlier era but which are now defunct, such as the Commonwealth Regional Development Program (1978-81). These graveyard policies continue to be administered in order to recoup loan repayments from firms assisted under regional schemes or are maintained on an "experimental" basis with reduced funding (for example, the Albury-Wodonga growth centre).

Secondly, there are regional measure that exist as a dependent adjunct of industry policy. This dependency is illustrated by the Steel Industry Package. The package began as a sectoral initiative, the Steel Industry Plan (SIP), and later had a Steel Region's Assistance Program (SRAP) tacked onto it. The relative achievements of the twin elements of the Steel Industry Package reflect the degree of commitment attached to Commonwealth regional policy. SIP has been remarkably successful in returning the industry to profitability whereas the SRAP is regarded as an unmitigated disaster by the regions it is designed to assist (Donaldson, 1985).

The low status of regional policy in Australia is further underscored by the government's 1984 administrative reorganization whereby the Labor government split regional and industry development functions into two separate federal departments. The regional function was assigned a less important portfolio signifying a downgrading in the importance of regional policy.

In Canada, a major government reorganization in 1982 effectively dismantled regional policy as an independent line department at the federal government level when the responsibility for industrial and regional policy was combined in a single department (Aucoin and Bakvis, 1984; 1985). Despite claims by government officials that the responsibility for regional policy had simply been reallocated, many analysts have argued that the change reflected a significant reduction in the importance of regional policy (Savoie, 1984; Lorimer, 1986). Moreover, a subsequent round of bureaucratic restructuring during 1987 has removed regional policy even further from the central channels of federal policymaking. The creation of a series of agencies responsible for regional policy in Atlantic Canada, Northern Ontario and the West has met with the criticism that regional policy runs the risk of being reduced to the level of local electoral and patronage politics (Globe and Mail, January 28, 1988, p. A4). Second, while a good deal of caution needs to be exercised in interpreting public policy expenditure data, recent trends indicate that resources allocated to regional policy are declining (Lithwick, 1986). At least five factors can be identified as significant in explaining the retreat of regional policy.

Changes in Public Policy Agendas

The dawn of the 1980s has witnessed significant shifts in public policy agendas in Australia and Canada. Both nations have become acutely aware of the extent to which internationalization has proceeded on the economic front and of the consequences for small open national economies. This realization has resulted in a reordering of priorities on the public policy agenda.

In Australia, not only is there no government policy or charter for regional development, but there now appears to be no self-evident case for a Commonwealth role in regional development in the 1980s. This is in part due to the findings of a whole series of Commonwealth-commissioned reports which gave credence to the need to adapt to externally induced change through policies to improve industrial structure and performance regardless of location (Jackson Report, 1975; Australia, 1977; Crawford Report, 1979; Myers Report, 1980). In fact, by 1980 the shift away from regional development policies and their replacement by national industrial strategies in the Commonwealth's consciousness was virtually complete.

These reports had considerable bearing on the evolution of contemporary industrial policy and established the pre-eminence of sectoral approaches in economic development. They progressively stressed efficiency objectives as the means of coming to terms with the structural pressures bearing on the national economy. Though the reports all noted (in varying degrees) the desirability of orderly regional adjustments to excess industrial capacity, all lacked specificity on how to achieve positive restructuring in problem regions.

In effect the reports sanctioned the subordination of equity considerations in policy formulation, and in so doing gave rise to the notion of a regional policy which acted as an adjunct of industry policy. This perspective is endorsed by the prevalent belief that inter-state disparities are not extreme, and that Australia faces no crisis of regional equity. Indeed, this belief is frequently vindicated by comparisons with the array of more serious regional problems confronting Canada (Higgins, 1981).

However, this view is open to question and, importantly, is disputed by evidence contained in a report from one of the Australian government's main research bodies. The report found that there is evidence of persistent and significant regional disparities at more disaggregated spatial levels (intra-state), that specific regions have been severely affected by structural change in recent years and that the observed regional disparities have tended to become entrenched (Bureau of Industry Economics, 1985).

Nevertheless, in spite of the evidence and despite calls from various state governments for improved co-operation on regional development policy (Cain, 1983; South Australian Government, 1984), in some cases emulating Canadian General Development Agreement arrangements, regional development in Australia is now seen as the dependent outcome of a national industry policy designed to promote positive but geographically indiscriminate change rather than as an initiatory mechanism underwriting the direction and scope of local change.

For Canada, the challenge of the 1980s has been described as one of "basically reworking our socio-economic structure in order to rekindle the failing engines of national economic growth and competitiveness" (Courchene and Melvin, 1986). The orientation toward efficiency and international competitiveness is clearly reflected in the Macdonald Commission report which has served to make bilateral free trade with the United States the dominant public policy issue in contemporary Canada (Canada, 1985a).

While the Macdonald Commission's mandate to examine the Canadian "economic union and development prospects" provided scope for investigating regional policy issues, the handling of these matters, as Wallace has noted in Chapter 14, proved vexatious to the enquiry (Bradfield, 1986; McNiven, 1987). The Commission's analysis was organized along economic and political fronts which appeared to proceed quite independently (Simeon, 1987). Whereas national economies were seen to be amenable to the Commission's neoclassical analysis, regional economies were not and were virtually ignored in the economic analysis. Regional policy was viewed almost exclusively as a political issue.

Moreover, from the Commission's perspective, regional policy could be seen as an impediment to negotiating a free trade agreement with the United States. Subsidization of production has been a point of contention in free trade discussions and American negotiators have stressed the need to establish a "level playing field" of continental proportions. The protectionist lobby in the United States has insisted upon reserving the right to apply "countervailing duties" to imported products deemed to have benefited from production subsidies. The significance of this threat has been made clear to Canadians in a number of recent trade disputes. That proponents of free trade want to avoid subsidization issues raised by regional policy is understandable.

It is clear that priorities on public policy agendas have been revamped in both countries. Australia's semi-peripheral position in the new international economic order has made deindustrialization a prime concern (Wilde, 1986). Coupled with a recent massive deterioration in the existing terms of trade, the viability of the whole economy has been brought into question. Under these circumstances traditional regional policy arguments favouring equity and redistribution are seen to be largely irrelevant. As a result, within the Commonwealth government, regional development is regarded by co-ordinating departments as not only unnecessary but also discriminatory, unconstitutional, too costly and contradictory to macroeconomic and sectoral objectives. The potential for region specific policies to improve aggregate growth, economic management and national welfare is largely ignored. Similarly, whereas the "Just Society" promoted by the Trudeau Liberals in Canada beginning in the late 1960s emphasized the importance of regional policy, efficiency and competitiveness have emerged as current priorities and regional concerns are not easily accommodated.

Analytical and Theoretical Ambiguity

Rational policymaking is normally seen to flow from an accepted analytical and theoretical framework. However, it is clear there is no agreement concerning the theoretical bases of regional development. In Australia, reports which have examined the regional impacts of structural change cite the absence of a theoretical framework as a problem for regional economic analysis (Bureau of Industry Economics, 1985; Logan, 1979; McLoughlin, 1986b). Similarly, the Macdonald Commission, having canvassed the research community, concluded "that relatively little is known about how and why regional economies grow" (Canada, 1985a, vol. 3, p. 198).

However, the neoclassical orientation of the Macdonald Commission's economic analysis needs to be emphasized (Norrie, 1986). While this framework occupies a dominant position in Canada, its utility for explaining regional development is being increasingly questioned (Bradfield,

1977 and 1986; Cannon, 1984). Alternative analytical frameworks are being used in examining the process of regional development. For example, some regional development research in Canada has begun to re-examine neglected indigenous theoretical frameworks such as Innis's staple theory, and to explore the rich tradition of Canadian political economy (Drache and Clement, 1986). Other work has adapted more general theoretical frameworks to regional analysis. For example, the world systems and NIDL literature have regional implications which are being explored (Weaver, 1985; Lipietz, 1985). Still others are investigating a wide variety of Marxist and neo-Marxist frameworks (Matthews, 1983; Clow, 1984). While this work has had little apparent impact on neoclassical analysis, there are indications that researchers within this latter tradition are cognizant of theoretical limitations and inconsistencies as well as practical problems with their own analytical framework (Polese, 1981; Vanderkamp, 1986).

The situation in Australia is not quite so progressive. While much of significance has been achieved by Australian industrial analysts, research has tended to concentrate on descriptive monitoring at the organizational and enterprise levels, and more recently at the sectoral and industry levels. Relatively little attention has been given to analyses based on the economic processes operating within regions themselves. Apart from the widespread application and refinement of the input-output technique as a tool of regional analysis (Jensen and West, 1985), the occasional incursion into regional economic determinism (Carter, 1983), and an exceptional attempt to reconcile regional and industry policy (Stilwell, 1985), the level of analytical and theoretical debate on regional economic development has been almost nil in the past decade.

There is one minor exception. The Commonwealth's Regional Development Branch is currently undertaking an Australia-wide analysis of regional economic structure and performance, based on the concept of regional comparative advantage. Couched strongly in neoclassical orthodoxy, the approach seeks to develop a regenerative rational for regional policy based on efficiency criteria (Department of Local Government and Administrative Services, 1987). However, to date there is no evidence to suggest that this approach has had any effect on government thinking or policy.

In Canada, the diversity of theoretical frameworks has created a rich intellectual climate for those involved in the study of regional development, whereas in Australia neglect and confusion are the case. Neither situation offers solace for the policy-maker who is confronted with the task of implementing policy today. In the absence of a broadly accepted theoretical framework policy frequently evolves in a very pragmatic fashion and has been subject to frequent changes in direction. These circumstances have had the effect of undermining the stature of regional policy when it is subjected to critical scrutiny.

Unspecified and Conflicting Policy Objectives

Lack of theoretical clarity has been paralleled by a failure to fully specify regional policy objectives. While there is no denying that governments have issued statements of principle concerning the need for policy to address regional balance and equity (Canada, 1985b) and to ensure that those most affected by structural adjustments do not bear a disproportionate share of the costs of economic change, the internal consistency of these principles and their compatibility with other policy objectives have often been ignored (Lithwick, 1982a). The failure to clarify and co-ordinate policy goals has resulted in a number of dilemmas on the regional policy front.

A particularly difficult problem has been that of deciding to what extent regional policy has developmental (efficiency) as opposed to compensatory (equity) goals. In both Australia and Canada, this discussion has crystallized around an elaborate system of regional and personal income transfers. Interregional income redistribution represents an attempt to compensate for unequal fiscal capacities of the Australian states and Canadian provinces (Economic Council of Canada, 1982; Courchene, 1984). Behind the practice lies the principle that Australians and Canadians, irrespective of place of residence, should be entitled to similar levels of public services (OECD, 1984). Income transfers to persons also result in some regional redistribution. For example, unemployment insurance payouts are biased toward regions with higher than average unemployment rates. Moreover, the Canadian program has a regional tilt in the sense that eligibility criteria and benefit levels are regionalized.

Critics argue that income redistribution and fiscal compensation programs have been pursued too vigorously. As a result, market adjustment processes are reduced and low income regions have fallen victim to a condition of "transfer dependency" (Courchene, 1978; Courchene and Melvin, 1986; Rabeau, 1987). Rather than representing a condition of disequilibrium, regional income disparities are seen to reflect a "policy induced equilibrium".

While the neoclassical basis of the transfer dependency argument has been strongly attacked in Canada (Matthews, 1981) and to a lesser extent in Australia (O'Connor, 1986), it is evident that income transfers are significantly larger than expenditures on programs which have a more clearly developmental orientation.

Institutional arrangements have also created potential for goal conflict in regional policy. In Canada, for example, federal manpower policy has always remained institutionally separate from regional policy (Smith, 1984). This creates a need for interdepartmental co-ordination of manpower training and mobility programs with regional job creation programs, a task not always achieved. The situation is similar in Australia where only one labour market program has a regional dimension (Labour Adjustment Training Arrangements) though future adult training schemes could become more regionally targeted (Kirby Report, 1985). Moreover, vigorous pursuit of national manpower mobility programs has always been constrained by the political realization that state and provincial governments would be unwilling to sanction large scale out-migration.

Finally, clarification of policy objectives becomes even more complex when the distinction between "explicit" and "implicit" regional policy is recognized. While a great deal of attention is focussed upon policies which originate with agencies and departments having an "explicit" regional development mandate, it is clearly evident that the impacts of general and sectoral policies are regionally differentiated. The magnitude of these "implicit" regional impacts may be very significant relative to those of "explicit" regional policy. For example, a variety of national policies has always been viewed as having uneven regional impacts (Armstrong and Taylor, 1978; Wadley, 1986). While the precise regional impacts of tariffs, monetary and fiscal policies, transport regulation and other national policies are difficult to specify (Whalley and Trela, 1986), on balance, they are generally considered to have favoured core as opposed to peripheral regions (Stilwell, 1983a) and the case is often made, at least in Canada, that regional policy is in part an attempt to compensate for these policies (Bercuson, 1977; Norrie and Percy, 1983).

In summary, the respective roles of the developmental and compensatory objectives of regional policy need to be clarified. Moreover, the implicit regional impacts of public policies must be better understood. The problem of achieving an appropriate institutional fit for regional policy contributes to these shortcomings. Defence of regional policy is made more difficult in the face of apparent policy contradictions and inconsistencies.

Results of Regional Policy

Flagging commitment to regional policy is in part attributable to difficulties in demonstrating that policy has produced significant results. While difficulties of evaluating regional policy are well known, consensus seems to be emerging on some issues. First, while results can vary tremendously from project to project, investment incentives provided to private business are viewed as the least successful dimension of Canadian and Australian regional policies (Borins, 1986; Canada, 1986; Piteman, 1982). Second, development agreements negotiated between the Canadian federal government and individual provinces have offered flexible instruments for tailoring policy to particular regional requirements. However, the comprehensive and long run nature of development associated with many of these projects has made evaluation problematic (Savoie, 1981 and 1986a). While there appears to be underlying agreement that low-key projects suited to local requirements have better prospects for success than many high profile mega-projects, difficulties in demonstrating the positive effects of such projects have weakened the political will to continue.

Third, some have claimed that regional policy has fragmented the national economy, weakening its capacity to respond to pressures for structural adjustment. In Australia it has been argued that regional policy may very well be responsible for "building-in" future regional adjustment problems (Industries Assistance Commission, 1981). The effect of these arguments is to further discredit regional policy.

On the positive side, income transfer programs have had an important effect in reducing provincial fiscal capacity differences. Thus regional disparities in levels of public services are no longer considered excessive (Courchene and Melvin, 1986; Lithwick, 1986). Paradoxically, the success of redistributive policies has eliminated the need to innovate further with them.

Regional Policy and the Politics of Federalism

Since the responsibility for regional development is not clearly assigned to a particular level of government, regional policy can be intensely political in federal states. As a consequence, regional policy can depend to an inordinate degree upon the ability to manage relations between respective levels of government.

In Canada, the formulation of regional policy has proceeded in a federal-provincial context. However, the financial strength of the federal government, in conjunction with its desire to address regional disparities from a national level, resulted in it assuming a leading role. At the same time, critics have argued that federalism has forced the federal government to deal with provinces on an individual basis, thereby effectively ruling out an approach which would emphasize the integration of regions within a national economy (Lithwick, 1982c). While it is debatable whether a more nationally integrative approach to regional policy would solve the problems of have-not regions, it is clear that the politics of federal-provincial relations has crucially influenced the form taken by a number of dimensions of Canadian regional policy.

First, regional development policies initiated in various federal government departments in the early 1960s were consolidated in a new Department of Regional Economic Expansion (DREE) by the end of that decade. This was seen to reflect the federal government's centralist approach to regional policy (Savoie, 1986a). Partly in reponse to criticism from the provinces, the federal government moved toward decentralization by introducing the General Development Agreement (GDA) in 1973 as its principal regional policy instrument. The GDA provided generous federal funding for projects initiated by the provinces and involved extensive decentralization of federal regional development machinery. However, the connection made in the public mind between regional policy and the provinces was unsatisfactory to the federal government. This resulted in the federal government introducing yet another policy instrument in 1982, the Economic and Regional Development Agreement (ERDA). Similar in substance to the GDA which it replaced, the ERDA is distinguished primarily by its centralized (federal) delivery system.

The system of equalization transfers also reflects the importance of federal-provincial relations to regional policy. While the principle of fiscal equalization may be the "glue that holds Confederation together", the level of equalization is subject to renegotiation every five years. Moreover, it seems clear from recent negotiations that the federal government, which holds the upper hand in these negotiations, is determined to reduce its level of transfers to provincial governments (Courchene, 1983).

Finally, whereas provinces are very dependent upon the federal purse in the case of explicit regional development and equalization policies, as owners of natural resources they are able to exercise significant control over the use of resources as instruments of "province building" (Tupper and Doern, 1981; Gunton and Richards, 1987; Richards and Pratt, 1979). The escalation of commodity prices in the 1970's made provinces increasingly aware of the magnitude of resource rents that could be tapped for provincial purposes. While federal-provincial conflict was most pronounced with respect to oil and natural gas, widespread use of resources as instruments of provincial development has constituted a challenge to the federal government's ability to pursue regional policy from a national perspective. Significant differences in provincial resource endowments in a period of violently fluctuating commodity prices were not only capable of markedly influencing levels of regional disparity but also wreaked havoc upon the federal fiscal equalization program (Scott, 1975). Thus, in a variety of ways regional development policy demonstrates the need for federal-provincial co-operation and harmonization.

In Australia the situation is perhaps less volatile but only insofar as regional policy has never had the political prominence it has had in Canada. The judiciary's interpretation of the Constitution (Stillwell, 1983b), the division of economic powers in the Australian federation (Groenewegen, 1983) and the fact that the 1977 White Paper on Manufacturing Industry more or less eschewed Commonwealth responsibility for regional initiatives on the grounds that they were the prerogative of the states (Wilson, 1978), all serve to restrict Commonwealth incursions into regional policy. Any Commonwealth attempts to direct regional development in Australia must therefore include state participation. However, the perpetual compromise between the need to implement regional policies with national objectives whilst preserving State autonomy over regional aims and interests presents a real problem. Several commonwealth strategies have been undermined in the past by the centralist tendencies of the different state governments, all of which pursue economic policies that reinforce the economic advantages of their metropolitan capitals at the expense of the regions (Howard, 1977).

This is particularly true of the two main Commonwealth forays into regional policy under the Ministry of Post War Reconstruction in the 1940's and the Department of Urban and Regional Development (DURD) in the 1970's (Coombs, 1981; Lloyd and Troy, 1984). Both experiences took place under Labour administrations and, as a result, the present Labor Government, conscious of past failures, is particularly sensitive. Moreover, this sensitivity extends to the federal bureaucracy where the "spectre of DURD" still looms large.

Apart from the fact that the bureaucracy lacks commitment to area-based policies, regional planning is also perceived as a threat to functional autonomy and authority. Powerful functional bureaucracies are difficult enough to co-ordinate at a single level of government, but when several tiers attempt to combine in the course of innovative policy development they prove unwieldy and act as an effective brake (Logan, 1979). In effect, intergovernmental relations and the nature of government apparatus play a critical role in undermining the potential of regional policy in Australia.

17.2 EMERGING POLICY DIRECTIONS

Public policy is shaped by both ideological and practical considerations. Current circumstances have created a distinct tension between these forces. Neo-conservatism emphasizing minimal state intervention and a preference for macro and framework policy instruments has come to dominate the ideology of policy-making (Gonick, 1987). However the logical implications of a philosophical position are not always translated into actual policy. First, it is necessary to separate neo-conservative rhetoric from reality. Observation of the performance of neo-conservative governments currently in power suggests that they are not unwilling to intervene in economic matters and that it may be more important to consider the nature than the extent of intervention (Clarkson, 1985; Magnusson, 1984).

Secondly, the effects of recent restructuring have been distributed very unevenly both sectorally and regionally. For example, economic growth during the current Canadian recovery has been narrowly focussed on industrial Southern Ontario and the peripheral regions, which appeared to benefit from the commodity boom of the 1970's, have languished. Thus, contemporary policy-making must attempt to resolve the dilemma created by the ideological impulse to defer to markets and political pressure to intervene to facilitate needed adjustments.

Consequences of Neo-conservatism for Regional Policy

The ascent of neo-conservatism has meant that pressure for structural change aimed at improving productivity and international competitiveness has become the single most important goal shaping Australian and Canadian public policy (Parry, 1982; Canada, 1985a). In Australia, policy is oriented to facilitating growth by shifting from a highly protected, insular economy to a much more exposed internationally integrated one. The nature and complexity of the factors generating economic change are seen to have reduced the efficacy of regional policy (OECD, 1983; Cameron and Houle, 1985; Weaver, 1985). At the same time the status of broad sectoral initiatives and industry policy has been elevated. As a result, regional policy has dropped out of the mainstream of national economic management. This thinking is summarized in the view,

> that regional policies have little relationship to overall national economic and social objectives, that they are narrow reactive policies whose sole purpose is to limit the local damage of sectoral adjustment to politically and socially acceptable levels (Hayes, 1986).

In Canada, gains in productivity and international competitiveness are to be sought primarily through further economic integration with the United States. The Macdonald Commission suggest that free trade "would make a major contribution to Canada's regional development and to national competitiveness and overall confidence" (Canada, 1985a, 1, p. 331). Continental free trade has been widely accepted among proponents of the neo-conservative paradigm as the basis of a solution to problems of national growth and regional disparities (Courchene and Melvin, 1986). While the subordination of regional policy is more explicit in Australia, the effect of the policy reorientation is no less striking in Canada. However the credibility of the proposed strategies needs to be examined.

In Australia, the preoccupation with aggregate and sectoral dimensions of economic policy has meant that suitable mechanisms to translate industry initiatives onto the ground (Steketee, 1986) and to deal with the regional implications of structural change have been ignored. In fact, it can be argued that industry policy would derive very real benefits from closer integration with regional policy.

First, by excluding regional interests from the development of industry strategies which affect their regions, the relevance and applicability of sectoral initiatives is seriously reduced (Donaldson, 1985). Conversely, where regional bodies are able to participate in the production of industry plans, the scene is set for industrial restructuring and regional adjustment to progress relatively smoothly (Lawrence, 1986).

Second, interdependence between different industries often leads to quite wide variations in production or employment growth of the same industry in different regions. It follows therefore that certain industries which may be "national losers" are actually "regional winners" and can contribute to overall growth in the economy. This feature tends to be ignored in purely sectoral policies. It becomes clear that:

> A successful national industry policy designed to improve productivity and employment must focus on the sectoral concerns of individual regions. A simplistic approach that concentrates on industries at the national level cannot be successful and may actually exacerbate regional problems (Bell and Lande, 1982).

Third, evidence supports the view that firms rather than industries should receive more attention in policies designed to facilitate structural adjustment (Bureau of Industry Economics, 1979). Without a regional dimension to industry policy, it is not possible to determine the nature of local inter-firm relationships and adjustment capabilities, nor offset some of the less savoury features of industrial restructuring which current industry plans fail to address (McLoughlin, 1985).

In Canada, the Science Council has summarized the need to co-ordinate industrial and regional policy in the following terms:

> It has become virtually impossible to discuss industrial policy in Canada without immediately addressing the problem of regional economic competition and its political manifestations federal-provincial and interprovincial conflict (Jenkins, 1983).

Moreover, the neo-conservative claim that free trade with the US will reduce Canadian regional disparities is not accepted by all analylists (McNiven, 1987; Bradfield, 1986; Savoie, 1986b). Historically economic integration achieved through free trade has tended to bestow the greatest benefits upon those interests and areas that begin from a position of relative strength. The experience of the European Economic Community, which is frequently cited as a model to which Canadians might aspire, only serves to demonstrate that whatever contribution economic integration can make to increasing efficiency and aggregate wealth, it is not a solution to the regional problem (Kiljunen, 1980).

Because of the strength of regionalism in the Canadian political fabric, it is understandable that regional policy has not been abandoned. On an official level there is an effort to convince that regional policy remains a national priority (Canada, 1985b). On a political level, recent years have witnessed a flurry of administrative and programmatic changes intended to reinforce this image. When regional and industrial policy were combined within the new Department of Regional Industrial Expansion (DRIE) in 1982, official statements stressed the increase in policy efficacy and administrative efficiency expected to result. However representatives of "have-not" regions did not accept the neo-conservative argument that "less is more" but saw the change as a double blow. Not only were total resources available to the policy envelope reduced, but a large proportion of the remaining funds were being diverted to central regions in support of various aspects of industrial policy.

To deflect this criticism, the government has announced a stream of new initiatives even as DRIE, the former flagship of regional policy continues to decline. Atlantic Canada has been the most obvious target for these initiatives with Enterprise Cape Breton, the Atlantic Enterprise Program, the Atlantic Enterprise Board and, most recently, the Atlantic Canada Opportunities Agency having been established in rapid succession since the beginning of the current Mulroney mandate. The regional agency model for delivering regional policy which was introduced in Atlantic Canada is being extended to Northern Ontario and the West. However, uncertainty surrounding the source of budgetary allocations for the new regional agencies coupled with the announcement that the regional development component of DRIE'S budget has been overcommitted only two months into the current fiscal year, leaves at issue the matter of the total financial commitment to regional policy. Moreover, the residual status of regional policy is further suggested with the shift of the industry policy mandate from DRIE to the Department of Industry, Science and Technology. This change appears to leave DRIE as an empty shell while emphasizing the technological dimensions of industrial policy.

Another strategy of the current government is to stress the "implicit" regional benefits accruing from more general government programs. For example, the termination of the National Energy Program ended the federal government's claim to sizable oil and gas revenues which now remain with producers and producing provinces. Secondly, the intent to extend the use of government procurement programs to serve a regional policy role has been announced. Finally, there is an effort to link the emerging interest in defence policy to regional policy. The allocation of defence contracts to firms in depressed regions is a strategy used widely in the United States. However, one study suggests that Canadian depressed regions are unlikely to have the industrial base to support substantial military procurement, and that the bulk of expenditures will eventually be felt in central Canada as a result of sub-contracting (Globe and Mail, June 5, 1987, p. A12). In summary, while there is an effort to draw attention to the regional benefits of a range of government initiatives, it is also clear that the policy pattern fits neatly into a neo-conservative agenda.

It is clear therefore that the responsibility for regional policy is now deemed the reserve of the Australian states and is being shifted to provincial and local governments in Canada. The Macdonald Commission argued that the federal government: should not involve itself directly in regional job creation. Its responsibilities end with its commitment to overcome regional productivity gaps and labour market imperfections" (Canada, 1985a, 3, p. 219).

The neo-conservative preference for markets implies a high priority for decentralization and privatization. However the limitations and inconsistencies inherent in such a strategy are well known (McNiven, 1987). More specifically the strategy can be seen to involve a subversion and co-option of the "small is beautiful" philosophy (Overton, forthcoming). Certainly the virtues of small enterprise and local initiative are currently held in high esteem (McLeod, 1987; Mansour, 1987). Furthermore, the contribution that such initiative can make to revitalizing a regional economy is not to be ignored. However, in an economy dominated by interregional capital, the limits of local initiative must be recognized. Peripheral regions in both Australia and Canada have long traditions of local initiative and self-help (Brym and Sacouman, 1979; Forbes, 1979; Local Government Ministers' Conference, 1987). That such activities were unable to solve local development problems in the past needs to be recognized before unreasonable expectations are placed upon "bottom-up" strategies.

In summary, it is apparent that neo-conservatism has become a predominant force shaping public policy in both nations. In Australia, regional policy has clearly been subordinated to national industrial policy. In Canada, industrial policy is pursued in the guise of free trade. The question of how industrial policy can be attuned to the regional effects of structural change goes largely unheeded. The question of whether regional policy can be something other than an adjunct and administrative afterthought of industry policy is currently being ignored.

Future Prospects for Regional Policy

Attempts to forecast how the tension between ideology and practical politics will be resolved in the realm of public policy are always highly speculative. However, the prospects for a resurgence of regional policy are slight as long as neo-conservative logic dominates government's approach to public policy. Two themes in neo-conservative thinking seem particularly important. First, regional policy is seen to be wholly aligned with equity and in conflict with efficiency considerations that are accorded higher priority. However there is considerable literature that argues that efficiency and equity need not be competitive and under certain conditions may be complementary.

Second, viewed from a neo-conservative perspective, policy intervention is justified primarily in response to market failure. However, the desire to solve the regional problem through a return to markets confronts the paradox that market failure did in fact constitute the original rationale for the introduction of regional policy.

Thus the theoretical basis of the conservative strategy appears flawed and there is little reason to expect the regional problem to subside. It is in this failure that the future of regional policy may lie. In countries as large and diverse as Australia and Canada, it is impossible to conceive of a single national approach to regional policy. Moreover, recent research in the field of development has provided greater insight into the way particular regions and communities are linked to the international economy. The vulnerability of regions and sectors to the vagaries of markets has become apparent, particularly in peripheral and semi-industrial economies. The hope for regional policy is that greater understanding of this reality will serve as a basis for tempering the excesses of the neo-conservative preoccupation with markets and stimulate innovative policy responses which will harness capital and labour to serve the interests of communities and regions as they strive to adjust to the emerging international economy.

18. INDUSTRIAL POLICY IN CANADA AND AUSTRALIA: TECHNOLOGICAL CHANGE AND SUPPORT FOR SMALL FIRMS

John N.H. Britton

This chapter compares the policies of Canada and Australia as they address the problems of manufacturing industry and examines in detail the industrial initiatives that have been taken to assist technological development, especially through the small firm sector. As a first step, the broad industrial, technological, and political pressures that bear on these two economies are reviewed because they set the agenda for the policies and programs that have been put in place.

Australia and Canada have not been successful in developing policies with significant impact on either the adverse effects of multinationals operating in the manufacturing sector or the widening technological gap between domestic and international industrial performance. It is widely recognized that foreign direct investment (FDI) in both countries has produced industrial jobs but has not established export capacity. Neither country has exploited the opportunities for development based on resource strength: high value-added exports of resource products are limited and neither country is a strong supplier of capital equipment. FDI has also generated a high propensity to import and as these imports are of increasing technological content, the lack of developmental stimulus associated with these leakages is of increasing significance. Furthermore, the pace of technological change, internationally, leaves host economies such as Australia and Canada vulnerable when economic activities (and job types) are allocated by multinational corporations: basic resource extraction, final assembly and local marketing are often the only functions available.

The trade balances in technologically sophisticated goods for the two countries are negative and worsening and, given the international context in which intra-corporate flows occur, the problems are large in relation to the power that these nations have to modify their trade patterns. Other factors are also important. First, policies are changed in accordance with perceived electoral attractiveness and with electoral results although the requirement for development is for long-term policy commitments. Second, there never has been a strong or consistent consensus identifying FDI or massive technology trading deficits as "problems." To a large degree this common characteristic may be attributable to the persistence of the "resource bail-out" myth — the ultimate component of the "staples trap" outlined by Watkins (1963). Third, there is marked dissonance between the voice of neoclassical economists advocating minimal intervention and the attempts of other bureaucrats, social scientists and many elected officials to identify alternative policy options. Fourth, the policies that have been implemented in Australia and Canada have been fashioned in a piecemeal fashion and do not represent clear strategies.

18.1 AUSTRALIA-CANADA ECONOMIC DIFFERENCES

The economies of Australia and Canada are similar enough in political-economic traditions, staples dependence, and administrative and economic structure, to invite direct comparisons. Yet there are significant differences which need to be made explicit. Thus the economic-size discrepancy between Australia and Canada is compounded by Canada's North American geography, its economic interpenetration with the US, and its greater manufacturing and service sector specialization. Traditionally, Canadian tariff levels have also been significantly lower than Australia which implies that Canada is better placed to change its industrial structure. Indeed, the high level of Australian tariffs which are more than twice the Canadian level (Olsen, 1982, p.134) constitute 90 percent of Australian assistance to manufacturing (Gregory, 1985). Protection in both countries is associated with labour-intensive production methods and a preponderance of small enterprises which have high levels of vertical integration, low levels of R & D, and poor exports. Conlon (1985) concludes, however, that Australian-Canadian industrial differences emerge most clearly in their trade patterns — Australian manufacturing being more obviously concerned with the export of natural resources while Canadian manufacturers are more technology intensive. That is, given that both countries are characterized by major deficits in their technological balance of payments, Canada, unlike Australia, has "significant" per capita technology-based exports: between 1978 and 1987, for example, Canada's R&D intensive exports increased from 7.8 to 10.0 percent.

Although more than 70 percent of Canada's trade is already with the US, concern that it was about to face considerably increased protectionism in US markets stimulated Canada's recent negotiation of a free trade agreement. This contains a sequential elimination of tariffs, some reduction in non-tariff barriers, and a dispute settlement mechanism to address issues such as the structure of industrial and regional subsidies. Clear results after a year are elusive though some unanticipated relationships have been observed — significant upward shifts in the exchange value of the Canadian dollar have thwarted access of Canadian producers to the US market. This short-term problem reflects major inflows of capital into Canada which derive from reassessments of the investment value of Canadian companies and from Canadian monetary policy. Other impacts such as withdrawal of branch plant operations from Canada and the capture of scale economies by Canadian producers which have been discussed in Canada require time to emerge. In this respect the 1983 trade agreement between Australia and New Zealand has been of interest especially as interpretations point to few adverse effects and greater benefits for the smaller trading partner (de Grandpre Report, 1989). Clearly, the trade agreement gives sharpened purpose to the structure of Canada's technological policies since they are both developmentally oriented and concerned with adjustment assistance for firms.

Canada has been successful in generating some technology-based industries such as telecommunications. Part of this pattern of industrial development involves the creation of spinoff firms from federal research establishments and from corporations that supply the regulated, domestic telecommunications service industry. In particular, there is a visibility to the joint public-corporate concentration in the telecommunication and computer fields in the Ottawa region (Doyle, 1986).

In the performance of medium-technology industries there are also contrasts between the two countries: the auto industry which is important to both provides a good case. Canada negotiated a lengthened lease on the auto industry through the Autopact of 1965. Admittedly, the jobs in assembly firms that have been retained are mainly in production but opportunities for innovation in auto parts are creating a broader mix of jobs. By contrast to new investment in Canada for the North American market by North American, Japanese, and South Korean firms, Australia has been rationalizing its auto industry to combat the pattern dating from the early 1960s of rising lines of protection, an increasing number of producers and declining levels of performance. The current thrust is to rebuild local design capability, to cut the number of models produced, to increase local value added, and to increase components exports (Australia, Department of Industry and Commerce, 1984).

Finally, Crown corporations in Canada, more so than in Australia, have provided a means whereby the limited scale of the national market might be minimized as a constraint on the ability of the economy to develop products for which there are international as well as domestic markets. As an outcome of the 1982 shift from Liberal to Conservative governments many of these enterprises have been sold to private interests together with Crown corporations that were formed when firms in danger of failure were "saved" for technological reasons. Thus de Haviland, Canadair, Connaught Laboratories and Air Canada have been sold to corporate buyers and privatized by public share offerings.

Both countries have moved to privatize, deregulate and increase competition by liberating the access of foreign financial institutions, industrial firms, and investment to domestic markets. Against the background of their economic differences Canada **may** fare better from the impact of neo-conservative policies sweeping through central government programs. Nevertheless, both these small, dependent, high-income economies must calculate their "benefits" from neo-conservatism quite differently from the large technology and capital exporters who retain control and innovation functions in their international corporations. It is important to recognize that the policies that are in effect in fully industrial economies such as the US and UK include interventionist choices that are clearly directed to assisting the growth of particular economic activities. By way of contrast and irony, the UK, despite its wave of privatization and deregulation, has introduced explicitly **active** industrial innovation programs that are successful in their industrial impact. These policy measures acknowledge that even "freed" markets require assistance in order to function efficiently. The last part of this chapter explores the logic of this approach, as part of an interpretation of recent research, because it is relevant for the development of policies for small firms in both Australia and Canada.

18.2 THE HOST ECONOMY PROBLEM AND SMALL FIRMS

How important is FDI in shaping the problem if not the policies for **domestic** industrial development? As outlined earlier, intra-corporate trade is an essential component of the market and technological success of multinational firms, and only the minority of these award world product mandates, or regional mandates, that permit significant local R & D and component contracting in Australia or Canada. In both countries there is substantial concern about the truncated form of industrialization, particularly the low level of R & D and the high level of technology imports, but the politics of government recognition of the negative implications of FDI have long been very complex given the high levels of foreign investment.

In Canada, the decade from the late 1960s to the late 1970s was one of serious debate about the impact of foreign ownership. Thus the Watkins Report (1968) was followed by the Gray Report (1972) which led to the establishment in 1974 of the Foreign Investment Review Agency (FIRA); a clear description of the high import propensity of foreign firms (Statistics Canada, 1978); and a technological assessment of the developmental implications of the high level of FDI (Britton and Gilmour, 1978).

The latter work analysed structural weaknesses in Canada's industrial development and focussed particularly on the negative technological consequences of the high level of US ownership. By way of prescription it explained and amplified the strategy of technological sovereignty. As a strategic option this had its clear alternative — free trade — propounded by the Economic Council (1975) and orthodox economists. Faced with these choices the Trudeau government retreated: the possibilities of an ambitious strategy could not outweigh the imputed political costs over the life of the administration. Thus, incrementalism prevailed as Canada pursued its defensive posture characterized by unco-ordinated policies (French 1980, pp. 130-1).

In Australia there was a comparable sequence of reports; the Jackson Report (Australia, 1975) led to the establishment in 1976 of a Foreign Investment Review Board to screen new foreign investments, and work by Crough and Wheelwright (1982) examined the extent to which Australia had become a "client state" through FDI. But neither country has shown a real inclination to reserve industrial assistance programs for domestic firms and in this respect the implications of foreign ownership have taken a back seat to the issue of unemployment. The recession certainly intensified this policy thrust and no government has shown itself willing to trade off "dubious" long term advantages for the short term social and political necessity of capturing jobs.

From literature and conversations it is clear that both Canadian and Australian officials are aware of policy developments in the other country, though Australia often seems to be able to evaluate Canadian ideas before deciding if policy commitments are warranted. This is illustrated in an interesting way by Australian reactions to studies of Canadian technological development: a discussion paper within Australia's Department of Industry, Technology, and Commerce (1985) lists nine propositions on the "truncation" of industrial structures derived from Gilmour (1982), and denies the applicability of any of these ideas on the impact of FDI to the Australian situation! By contrast, more positive public discussion in Australia linked the Canadian arguments on technological dependence to Australia's poor R & D (input) and innovation (output) performance in manufacturing (Morris, 1983; Grant, 1983), and in official publications in 1985 the relatively low level of R & D in Australia by multinational corporations is clearly noted (Australian Departments of Science, and Industry, Technology and Commerce, 1985).

The connection between FDI and technological underdevelopment clearly has been lost on some in Canberra's bureaucracy; nevertheless, dissembling has not been the style of Barry Jones, Australia's Minister for Science and Technology, who clearly identified his position as follows: "The phenomenon of Australians lending money to multinationals to enable them to acquire control of the Australian economy... confirms the impotence of Australia's political system, and the failure of its governments to grasp the nature and extent of economic and technological change" (Jones, 1983, p. 223). As though wishing to confirm Jones' argument, the Labor Government in Australia adopted measures in 1986 to relax foreign investment guidelines to the extent that investment proposals are automatically approved by FIRB unless judged contrary to the national interest. This shift in Australian policy which had been urged by economists (Kasper, 1984), as well as by bureaucrats, echoed a major shift in Canadian policy under a Conservative government which in 1984 dramatically eliminated FIRA, replaced it with

Investment Canada (Safarian, 1985) and began actively seeking foreign investment as promised in its election platform. While "outside control fell to 48 percent from 61 percent" in manufacturing in Canada between 1970 and 1985 (Macleans, May 4, 1987, p. 39) increases in foreign ownership have resulted from Canada's renewed openness to foreign investment and the more effective integration of North American capital markets. In 1988, for example, net capital inflows reached $4.9 billion as part of a positive wave which started in 1986.

Clearly there is to be no resolution of the debate on foreign ownership; governments have had to respond in the post-recession years to the need for job generation and for governments influenced by neo-conservative ideas this has meant fewer impediments or irritations for multinationals in their shifting of activities and investments on a world basis. Canadian and Australian firms are vulnerable to both plant closures as part of larger schemes of corporate rationalization and targets for takeovers. Canada's free trade agreement with the US, as Wallace explains in Chapter 14, now implies that some Canadian branch plants are more vulnerable to closure than ever before. Simultaneously, many Canadian high technology and resource companies are attractive investment targets.

18.3 TECHNOLOGICAL DEVELOPMENT

Although Canada has a superior record of industrial innovation and more R & D activity than Australia, it has itself fallen seriously behind comparable nations (such as Sweden) in industrial R & D. Nevertheless both Canada and Australia are included in the same group of countries by OECD and both are clearly in the "cellar" of this particular league (Table 18.1).

Understandably, over the past two decades, technological development has been a growing policy area in both Canada and Australia and although Canadian experience is by no means an appropriate Australian model, frequent reference in that direction is made in Australian reports. Both countries recognized the importance of public laboratories many decades ago — especially to service agricultural activities — but with a few exceptions suchas telecommunications in Canada, both have experienced some dissatisfaction with the efficiency with which technology is transferred to the industrial sector (Cordell and Gilmour, 1976).

TABLE 18.1

R & D PERFORMANCE-SELECTED COUNTRIES 1986

	total R & D/GDP Percent	industrial R & D/GDP Percent	Industrial R & D Total R & D Percent
United States	2.89	2.07	71.9
Japan	2.81	1.88	66.8
Sweden	2.79	1.97	70.7
Germany	2.66	1.92	72.2
France	2.38	1.40	58.7
United Kingdom	2.33	1.47	63.1
Switzerland	2.28	1.69	74.3
Netherlands	2.06	1.16	56.5
Italy	1.47	0.85	56.9
Canada	**1.35**	**0.68**	**50.8**
Australia	**1.14**	**0.39**	**33.9**
Spain	0.48	0.28	57.5

Source: OECD, Selected Science and Technology Indicators, Recent Results" in Ontario, MITT, *A Commitment to Research and Development: An Action Plan*, 1988.

In the late 1960s both countries developed science and technology policy units; the Science Council of Canada acts as a public voice with a Minister of State as a governmental reporting line, while the Australian Science and Technology Council directly advises the Prime Minister. In 1984 the Australian government moved the responsibility for technology from the Department of Science and Technology to the Department of Technology and this more central role to technological policy has been repeated in Canada with the recent formation of a similar department — Industry, Science and Technology. It is probably a more telling factor, however, that neither country has proven itself capable of specifying its technological priorities and its strategy for realizing this goal. In Canada there have been abortive attempts to identify a national technological strategy before the Free Trade Agreement (French, 1980) and after (Atkinson and Coleman, 1989) as a policy response to adjustment problems. Australia got to the stage of printing a Revised Discussion Draft of a Technology Strategy (Australia, Departments of Science, and Industry, Technology and Commerce, 1985) but the consultative process ground to a halt as Australia experienced difficulty under its post-1983 Labor government of matching actual strategy and policies for technological development with interventionist rhetoric (Joseph, 1984). Both countries have programs that cover obvious industrial policy areas — for example, labour adjustment assistance for declining industries and subsidies for R & D costs of firms — but neither has approached planning the funding of its universities as though their role in generating continuing technology growth is fully understood.

In Canada, R & D assistance occurred through a number of agencies including (1983 to 1987) the Department of Regional Industrial Expansion (DRIE). Until very recently, therefore, the main distinction to be made between Canadian and Australian programs for the technological assistance of firms was the way regional and industrial policy was confounded in Canada. The regional policy response of the Canadian federal government is complex since it involves provincial income equalization, payments from tax revenues which influence infrastructural expenditures, family income stabilization measures which operate through unemployment insurance and program funds for post-secondary education and wealth. In addition, DRIE functioned in its industrial assistance role through a locationally weighted program of discrimination against metropolitan centres. DRIE's grant and loan operations thus institutionalized an inverse interpretation of our understanding of how agglomeration processes influence higher rates of firm formation, survival, growth and technology acquisition (Britton and Gertler, 1986). In contrast, much evidence points to the implementation of Australian industrial policy being locationally neutral; for example, the way the Australian Industrial Research and Development Incentives Scheme has operated its grant program (Department of Industry, Technology and Commerce 1985).

Canada funds additional technological programs:

(1) The National Research Council (NRC) provides grants through its R & D (including engineering) mandate and has assisted small firms with the costs of new product development and with laboratory work. NRC also acts to transfer technology from its own labs to commercial firms, and in that respect has an active mode of program delivery.

(2) Other federal programs such as defence procurement and the Defence Industries Productivity Program tend to have some regional development aspects.

(3) There are attempts made to turn public sector research bodies into research contractors, diffusion agencies, and initiators of commercial research. (In Australia, an interesting example of the latter trend is SIROTECH, a commercial company formed by CSIRO to contract research to industrial firms; it generally facilitates the transfer of CSIRO research results to industry).

(4) Various federal government programs support strategic technologies, specifically, information technology (including the development and adoption of systems using micro-electronics), biotechnology, and advanced industrial materials, which are regarded as generic technologies. Industry-wide knowledge is generated through particular projects in individual firms.

(5) Technology diffusion is funded by means of diffusion centres; innovation centres have been established at the University of Waterloo (Ontario) and Ecole Polytechnic, Montreal (Quebec) in order that potential industrial innovations might be assessed. Most universities in Canada (and Australia) maintain offices that license technology and/or connect firms with research teams.

(6) Various types of assistance to help firms identify international markets have been inroduced in Canada (and Australia). Fundamentally, these schemes focus on "high technology" companies although there is a progressive development of data bases of all firms interested in exporting particular products or supplying them to a wider domestic market. The international marketing schemes provide financial assistance on a shared-risk basis, insurance, and financial guarantees to foreign buyers.

(7) Canada has embraced tax write-offs in a substantial way (Bernstein 1985). One scheme seriously backfired since tax concessions were traded

between firms, and R&D performance did not match the ability with which this type of paper entrepreneurism was practised. Nevertheless, a new generation of tax incentive schemes subsidize R & D and operate mainly in a locationally neutral fashion. Tax write-offs have been of increasing importance in Canada's R & D support portfolio: they are at least 44 percent of total R&D support by government. As might be expected the advantage of this scheme is bureaucratic, its administration is simple, and it accords with the general thrust of conservative policies to simplify industry-government relationships. Following Canada's experiments, Australia adopted a comparable scheme but one which incorporates direct grant support for initiatives by small firms. Since 1985, however, there has been a Canadian response to the financial problems of small firms and a system of refundable tax credits (not just write-offs) introduced which make payments to non-taxable firms undertaking R & D (Bernstein 1985). There has been a substantial increase in the rate at which this program has been used by Canadian firms — $374 million R&D tax credits in 1984 increased to $729 million in 1987.

18.4 THEORY

The measures to directly assist the amount of R&D undertaken by industrial firms rely on two principles. First, the linear model of innovation (the technology-push variant) as used over the past two decades suggests that a new product emerges as a result of a sequence of stages involving basic research (mainly in universities and public and private sector laboratories); applied research in firms (that attempts to resolve problems or to use the results of basic research); developmental experiments concerned with producing prototypes of new processes and products or systems; commercial manufacturing; and finally marketing. The importance of this conception of the process of innovation owes much to the public policies that have been put in place to encourage technological change because programs have been developed in most economies to encourage research and, it is hoped, industrial innovation and economic growth. Second, the legitimacy of government assistance for the private sector's R&D effort has been enhanced by the development of economic theory related to the possibility of "market failure." The work of Arrow (1962) pointed to the importance of a triad of factors that may impede the level of research that is industrially relevant: **indivisibility** - the high cost of research in relation to the size of individual firms, or even an industry; **inappropriability** - the benefits of research are usually not completely captured by the firm making the expenditure; and **uncertainty** - the high risk of failure in research. These factors individually and collectively warrant off-setting action by government. Australian and Canadian policies are aligned with this philosophy of R&D support and, more particularly, most expenditures to support innovation are directed to basic and to applied research in the public and private sectors in the belief that by increasing research input to the innovation process technological advance and industrial competitiveness will be enhanced.

Some problems with the theoretical base for R&D subsidies are recognized in Canada and Australia: the criteria for identifying market failure have been the subject of some debate (Ronayne, 1984) and the Australian Departments of Finance and Treasury have suggested that the assumption that there is market failure rather than acceptable risk should be disputed (Joseph and Johnston, 1985); that the R&D necessary to generate and to supply technology be treated as "simply another input...that...should be governed by market forces." The OECD finds this preposterous, using "quixotic" to describe such a view of the circumstances where market failure is rampant domestically and internationally (OECD, 1985). In Canada, some economists have expressed reservations about programs which "pick winners" and attention has also been directed to the problem of administrative quality in delivering appropriate levels of assistance (Tarasofsky, 1984).

In Freeman's conception of the innovation process, R&D is associated to an increasing degree with each of three scales of innovative activity (Freeman 1986): **incremental innovation**, a continuous process leading to a steady improvement in the array of existing products and services; **radical innovation**, major events which may lead to serious adjustments for firms in a particular industry as the diffusion process takes effect; **technological revolutions**, the formation of new sectors of the economy which power other changes of an incremental and radical type.

Both Freeman's ideas and empirical work show the practical importance of incremental forms of industrial innovation which are often identified with improved product performance and thence the retention or growth of a market position. Incremental innovation, therefore, exemplifies the coupling model of innovation and the importance of redesign (Rothwell, 1986). First, it stimulates the increased importance of engineers, designers, and production supervisors, rather than scientists (Gannicott, 1982; Pavitt and Walker, 1976). Second, it illustrates the fundamentally non-linear nature of the innovation process with its feedback mechanisms which closely link management, market, technical, design, engineering and research activities. Under the policy regimes of Canada and Australia, these aspects of the innovation process tend to conceptually separate from R&D and R&D assistance. One of the reasons may be the different degrees of access of small versus large firms to these inputs. The superior internal resources of large firms mean no problems are encountered: small enterprises, however, present a unique set of policy triads.

18.5 SMALL FIRMS

In the context of the high levels of FDI and small numbers of large domestically owned firms, industrial development policies which assist the survival and growth (even birth) of **domestic** small - and medium-sized enterprises are of considerable interest.

Various arguments are used to justify policies to support firms in the under-200 employees range. At least in Canada, small firms in manufacturing held jobs during the recession and were responsible for almost all net new job creation. This defence, though socially and politically strong, overlooks the economic benefits that may derive from the greater efficiency of larger enterprises and their international growth prospects. A further political argument suggests that it is prudent to provide assistance to small domestic firms which are financially constrained but which represent "entrepreneurship." In addition, it is recognized that small firms have been notable in the high technology growth experience of some regions in the US and the UK and if "appropriate policies" are implemented, a local version of technology-based growth is often thought possible.

Principles of Policy Design

Some differences in the policy needs of small versus large firms have been outlined, and discussion has shown the direction that policy initiatives might take. Nevertheless, small firms are not homogeneous in their needs and three distinctive types of innovative small-manufacturing firms may be recognized on the basis of different paths to product development; each has different requirements for technical inputs. These generate distinctive patterns of interdependence among firms and ultimately they call for different policy initiatives (see Rothwell and Zegfeld, 1985; Freeman, 1986; Britton, 1989b).

Traditional Industries with Specialist Market Niches

Firms in textiles, clothing and furniture can compete with imports from low-wage economies if they focus on well-defined market niches that exist for particular intermediate goods such as industrial textiles, that are specified in terms of performance standards or which meet design criteria. Input requirements for successful firms, therefore, include production and inventory efficiency, knowledge of new materials and markets, quality control, and the ability to implement performance and design innovations which can be obtained through contracts with professional consultants.

Modern Industries with Specialist Market Niches

Firms in scientific instruments and specialized machinery production, for example, are characterized by an **in-house** talent in technical skills (Rothwell et al. 1983) as a requirement for success. This core of skills is often associated with the founding of these firms and is required in order to make use of professional inputs from outside which maintain the range and quality of technical information. Competitiveness in these industries is related to low-price factors and to incremental design innovations rather than to the development of new technology.

New Technology Based Firms

These firms depend on in-house scientific design and engineering capability, which provide an intellectual integrity to the core of the enterprise which is often established in a prior "soft" phase of consulting work (Bullock, 1983). Enterprises producing electronic devices, CAD software and systems, or supplying innovations using advances in biotechnology, are good examples of this type of firm. Although they may contract-in highly specialized product testing, software services, or even certain types of design work, firms of this type are more likely to contract-out production activity in order to minimize their need for fixed capital and in order to stay close in function to the original conception of the firm (Segal, Quince, Wicksteed, 1985).

It will be clear from this threefold division of small enterprises that policies must recognize several different characteristics and relationships. First, only a limited population of small firms is R&D intensive and R&D programs will only apply to this few. Second, there are many actual and potential **innovators** who do not perform R&D, and increasing their performance requires a policy rationale and delivery appropriate to the incremental innovation case outlined earlier.

These and other policy initiatives that are applied to Australian and Canadian small firms are examined below.

Advisory programs

Although the declining cost of computer-based equipment allows small firms in secondary manufacturing to be very flexible in production, they are limited in the range of their technical capabilities and business information. In most cases this structural weakness could be offset through interdependence with other small firms, but there are search and information costs associated with the development of such linkages, and with arranging contracts. These can have a significant negative impact on the transfer of all forms of information to small manufacturing firms and it is critical to understand that when the acquisition of new information is impeded (by transaction costs) the rate of incremental innovation is retarded. If these market imperfections, with their dampening influence on innovation, are to be offset, active measures are required rather than passive or reactive policies. Thus if firms are inhibited in their interaction with other businesses because of search and information costs, then these same factors are bound to prevent them from taking advantage of government programs which do not actively seek needy firms. While this problem has been recognized by means of networking initiatives that have been designed in Australia, industrial policies in Canada make greater use of a passive mode of program delivery. Compared with the highly active interventionist approaches that can be devised and that are employed in the UK and in western Europe (Britton, 1989a; Sweeney, 1987), this is a surprising limitation on the practical options available to Canadian firms.

In Canada, technical advice for small firms is available through an NRC program which links its own technology advisors and agents drawn from provincial research organizations such as Ontario's ORTECH university technology transfer offices and technology centres of various types.

The relationship between NRC and provincial organizations has a long and successful record, but assesssments have indicated that NRC's technical advice programs are underfunded (Leroy and Dufour, 1983). ORTECH, founded in 1928, represents a major initiative in its own right, undertaking contract research and testing for firms of all sizes, but predominantly for small firms. SIROTECH, in Australia, is increasing its comparable functions but there are no state-funded research organizations like those of Canadian provinces.

The Technology Transfer Council (TTC) in Australia is one of a variety of technical support and advisory agencies which have emerged in recent years (1980) with Australian Commonwealth and private financial support, the latter including fees for service to firms and state agencies. The TTC, a responsibility of the Confederation of Australian Industry and the Metals Trades Industry Association, demonstrates to firms the value of business planning, the "just-in-time" production strategy, and quality control. The TTC, the Productivity Promotion Council, and the Industrial Design Council of Australia, which operates a design referral service and counsels firms on strategic planning and implementation and new product management, represent innovative national joint public/private sector institutions to provide technical advisory services. They do not have Canadian counterparts (Britton, 1989a) because they are (pro-)active organizations which seek clients otherwise unaware of the potential value of their service.

R&D

Both Australia and Canada operate a mix of loan and subsidy programs for R&D, as noted previously. In Canada, the Federal Business Development Bank (FDBD) is the public agency responsible for loans of last resort, while NRC delivers small firm research subsidies. The Canadian taxation definition of R&D, though recently revised, has a strong research thrust, and in this respect the language used by Australia appears to be encouraging firms to explore the breadth of assistance available. In its introduction of a 150 percent tax concession, for example, R&D is defined in Australia to include the cost of producing commercial computer software, industrial and engineering design, production engineering, operations research, and the design, construction and operation of prototypes. Access is provided to direct subsidies for (small) firms whose tax liabilities do not enable them to take advantage of these tax concessions. While this arrangement parallels Canadian programs, the stronger central co-ordination of innovation policies in Australia appears to have benefit for small- and medium-sized firms. In Canada the policy context is complex, since provincial governments also provide competitive tax concessions for R&D, especially incremental R&D.

Capital:

There has been a thrust to encourage the expansion of technology-intensive small firms and to overcome barriers which affect their growth such as access to equity capital. The provision of tax breaks for individual and corporate investors in venture capital funds has been the most recent policy change, with Australia following Canada's lead in this respect (Espie Report, 1983). Australia, however, now has well-developed "second board" share trading in junior high technology stocks which supplements the highly insititutional pattern of subscription to Management Investment Corporations which have tended to be conservative in their investment decisions. Over-the-counter stock trading in Canada has increased in recent years with the development of COATS (Canadian Over-The-Counter Automated Trading System). The need to provide greater access for start-up terms to pre-venture capital is generally recognized — perhaps the expansion of Ontario's COIN network (Computerized Ontario Investment Network) will assist, as will recent Ontario government initiatives.

Incubation:

The limited number of firms with special access to venture capital may be a serious limitation on technological development and so in Canada — and even more so in Australia — there have been initiatives explicitly concerned with the birth and incubation of new firms, especially small technology-based enterprises (Joseph, 1989). Most attempts to produce the necessary cluster of small firms are made concrete by state, provincial or local investments in incubators, and science and technology parks (Macdonald, 1983; Searl, 1985; Steed and deGenova, 1983) which are developed in the belief that high technology environments can be built using small, new, domestic firms, provided that public action helps to offset the disadvantage of small initial size and institutional barriers to growth (e.g. capital). These programs have a clouded record, especially as it is generally thought that in new firm generation the role of large firms is important. International evidence shows that science parks localize existing firms (of all sizes) seeking superior accommodation (Joseph, 1989) but only in exceptionally well managed facilities is there much evidence to show that they actually generate or enhance the inputs of many new enterprises. Associated with these ventures are attempts to encourage universities and colleges to produce entrepreneurs and firms to commercialize technological advances: these have local characteristics but are part of an international trend toward increasing the applied side of basic research conducted within tertiary education (see below).

State-provincial vs. national initiatives:

Given the familiarity each country has with the policies of the other, it is not surprising that there is a basic similarity in the technology policies of Canada and Australia. There are differences, however, in the emergence of policies at the national or state-provincial level reflecting the way responsibilities and funding have evolved in the two federal systems. Tax relief on venture funds for example, is an area of provincial experimentation in Canada, while the Commonwealth government is the main actor in Australia. The policy approaches at the provincial-state level have evolved qualitative differences, too, despite common concerns — here the policies of Victoria and Ontario are considered as they impact on the small-firm sector.

Three impressions dominate the Victoria-Ontario comparision. First, Ontario's concern is to generate strong connections between well-established technologically advanced firms and universities, and it has devised a number of programs to achieve an increased level of pre-competitive research undertaken as joint projects: over $60 million, 1988-89. Small technology-based enterprises are expected to spinoff from larger firms enriched by these policies. Pre-venture support for small innovative firms is allocated (an equity investment-and-recovery program) but Ontario is quite deliberate in its focus on the research and capital needs of threshold firms ($40 million plus revenue) and larger enterprises. Victoria also wishes to create "sunrise" or highly modernistic industries (OECD, 1985; Macdonald, 1983) but its thrust is to assist the generation of new technology-based firms through technology incubators (Ministry of Education, Victoria, 1988).

Second, both Victoria and Ontario maintained a strong non-metropolitan bias in their economic development policies until recently. The 1984 Labor government in Victoria reversed the previous arrangement and began to reflect its metropolitan power base by designing the locational content of its policies to respond to the technological needs of States and the geography of its industrial potential. Melbourne, for example, has a set of metropolitan advantages favouring the emergence and growth of new, progressive small firms. In this respect Melbourne and Toronto have similar advantages. By comparision with Victoria, however, Ontario saw a majority Liberal government enter office in 1986 and it is still uncertain how to both encourage growth throughout the university urban systems (12 non-metropolitan universities) and to capitalize on the metropolitan strengths of Toronto.

Third, a detailed comparison of Victoria's and Ontario's policies confirms the greater co-ordination efforts in Victoria especially on the "supply side" towards improving the access to information inputs, such as design, engineering, management, marketing and other services, that can be obtained by small firms on a consultancy basis. These differences imply a more comprehensive approach in Victoria compared with Ontario's greater policy specialization on technology development - these reflect the stronger development of Ontario's secondary manufacturing, the presence of a small number of threshold firms and the province's greater research infrastructure. They also reflect the clearly more conservative style of Ontario's programs. The result in Victoria is that programs give considerable recognition to the need for producer service intermediaries, and state support is given. By way of contrast, and as noted earlier, Ontario has neither private sector nor public intermediaries to assist small firms in a (pro-)active manner. It is perhaps symptomatic of this policy gap in Ontario that financial support has been shifted from R&D and technology diffusion (e.g. technology centres) to R&D activity alone.

18.6 CONCLUSION

It has been argued in this chapter that the industrial development of Australia and Canada have been inhibited by structural factors, especially FDI. In both countries domestic firms, particularly small enterprises, function in the shadow of massive imports by multinationals which maintain truncated operations especially in terms of their investment in innovation. The public policy actions that are potentially usefull focus on domestic firms, given that the present governments of Australia and Canada have no liking for actions that place them in conflict with foreign firms that may be planning new capital inflows and the expansion of jobs. Over the long-term, two policy options may be followed to increase the size and improve the vitality and competitive position of the domestically owned industrial sector.

First, on the **demand** side, the access of domestic firms to government procurement can be maximized. This implies that active demand management should be practised in all areas of public sector purchasing. It is clear from the international literature that the developed industrial economies pursue the advantages of this strategy — for example, in defence procurement and

R&D contracting in the US. But it is equally true that no level of government in Canada or Australia has yet mastered the management of the developmental lead times that this option requires if it is to be implemented at even the modest scale which would apply in the case of the Australian and Canadian economies. Although passing reference has been made to the demand-side approach in this chapter further policy evaluation would first require more action.

Second, on the **supply** side, both countries have been much more interested in attempting to remove obstacles — particularly market imperfections — to the efficient and innovative functioning of industrial firms, and this has been the focus of discussion here.

It has been acknowledged that demand- and supply-side issues are intertwined (in design functions, for example) but special attention has been given to R&D and other programs designed to increase the development and use of new technology by industrial firms. Both Australia and Canada have experimented with tax and grant programs aimed at increasing the output of commercially successful innovations. Given the high levels of technology imports by both countries, export successes registering new trading specializations are usually sought. With this goal in mind, one of the increasingly important areas of policy experimentation underway is the linkage of large- and medium-sized firms with university research teams. Attention has also been given to the generation of new technology-based firms which, though small ventures in their early years, are innovators and are regarded as investment opportunities for future growth.

Other forms of industrial innovation among small firms have been slower to receive the favour of program expenditures: apparently, a lower value is placed on incremental rather than radical forms of innovation even though pre-empting imports (or import replacement) and preserving jobs are significant for both economies. The social science literature on innovation, however, is much more realistic about the importance of firms which are active in the refinement of their production systems and which consistently strive for product improvements through redesign. The theory which underpins policies to improve the rate of incremental innovation among small firms has been given particular scrutiny here, as have the different policy approaches to this issue which have been evolving in Australia and Canada. Discussion has focussed on those initiatives which improve the access of firms to inputs of technical and business information; therefore, transaction cost problems have been examined as a way of identifying the imperfections of information markets that small firms must overcome if they are to enjoy the advantages of efficient producer-service networks.

Within this general approach to policy making it has been argued that governments should support the development of active programs in addition to environmental and essentially passive initiatives. The logic here is that if appropriate policies facilitate the operation of information markets, small firms in manufacturing and consultant firms in producer services will have a better chance of establishing creative industrial linkages; they will have achieved some success in overcoming the problems of learning about the range of services available and/or needed, the specific opportunities to obtain and supply information, and about negotiating contracts. By extension, the future viability of small firms in the manufacturing sector of these economies is closely bound up with the expansion and development of the producer service sector that provides necessary technical consulting and other informational inputs.

ACKNOWLEDGMENT

This research was supported by the Social Science and Humanities Research Council of Canada and is gratefully acknowledged.

19. CONCLUDING STATEMENT

Roger Hayter and Peter D. Wilde

An important assumption, albeit an implicit one, underlying the workshop upon which this volume is based is that nation-states are valid units of investigation and that Australia and Canada, in particular, comprise internally coherent politically functioning regions or, to use the term recently preferred by Scott and Storper (1986), "territories." In practice, as the contents of this book reveal, and as was made abundantly clear during the proceedings of the workshop, this may be the case, although increasingly powerful global processes are persistently undermining the assumption. In particular there is concern about the effectiveness of Australia and Canada, as medium-sized nation-states, in implementing policies to direct their economies.

Of course, questions concerning the viability of nation states, and of nationalism, in the arena of political economy are not new, and criticism has stemmed from various perspectives. In Canada, for example, the most critical analysis of the viability of nationalism is in writing in the neoclassical tradition (Watkins, 1978), while in Australia structuralist or neo-Marxian perspectives are usually most explicit! In essence, those individuals, whether conservative or radical who stress (with or without approval) that relatively smaller countries have substantially lost their discretionary powers over economic matters over the past several decades, point to the growing interdependence of the global economy, especially as reflected in the size and scope of multinational corporate operations. There is no doubt that recent trends towards an integrated global capitalist economy have complicated and restricted the means by which nation states can control their own economies and manoeuvre against each other. Indeed, an important thrust of this book has been to document the tensions and conflicts that have arisen from globalization for Australia and Canada, as two small or medium-sized and relatively open economies.

The view that smaller countries in general, and Australia and Canada in particular, can function only at the discretion of the major political powers or elements of capital, or from some global imperatives, as some would suggest (Vernon, 1977, p. 190) is too extreme. In this regard, Sayer (1986, p.108) is probably correct in arguing that the role of governments in shaping economic events and in bargaining with multinationals is typically underestimated (and not only by radicals!) After all, governments continue to have a wide range of policies available to them which in one way or another affect the structure of industry and the ability of industry to compete internationally. Many governments, most certainly including those of Australia and Canada, also have access to considerable financial and human resources and to a variety of infrastructure and natural resources. Governments, in other words, can make choices. For Canada, as Angers noted some time ago, the key choice is whether "to accept the rules of the world economy or to specify ... her own strategies within the world economy" (Angers, 1972, p. 131). Exactly the same point has been made with reference to Australia (Crough, and Wheelwright, 1982).

If the precise extent of the bargaining power enjoyed by Australia and Canada is unclear, it is generally accepted that Australia and Canada have largely preferred to accept "the rules of the world economy" with respect to trade, investment and technology and have foregone alternatives which would have given them greater control over their national economies and global roles. Moreover, it is argued by many, including several contributors to this book, that the limits to discretionary behaviour imposed by this "internationalist strategy" compound over time. Certainly, as a result of this strategy, the global roles of Australia and Canada, as indicated by export structures, have remained remarkably specialized over long periods of time. Indeed, the stability of their export structures stands in sharp contrast to the experience of most other advanced and rapidly industrializing countries.

If the internationalist strategy adopted by Australia and Canada has been very strongly associated with a highly specialized export role why have Australia and Canada remained so committed to such a strategy? This question would seem to be a legitimate one, since governments at varying levels in both countries have periodically expressed interest in trade diversification, and the desire for such a development is particularly pronounced at the present time. Conventional answers to this question might well emphasize the following points. First, both countries are rich in resources which have been in demand and whose exploitation in the past has permitted high levels of population and income growth; the need until now at least for trade diversification has simply not been pressing. Second, both countries were developed as colonies of the UK and subsequently willingly accepted the free trade doctrine promoted by the UK and the US. Third, over time, and for a variety of interrelated reasons, both countries fell into a "staple trap" which is continually re-created by the specialized nature of institutional and infrastructural investments, and by established free trade attitudes.

There may well be a fourth reason which has limited the abilities of the Australian and Canadian governments to change their global roles. This reason relates to the federal structure of both countries and the amount of political energy that is spent on domestic matters. Certainly, a recurring theme throughout this book has been the nature and significance of interstate and interprovincial rivalries and conflicts and the difficulties of achieving national consensus, even among governments. It seems at least reasonable to speculate that so much political investment in both countries is spent on seeking to resolve domestic issues, many of which directly impinge upon interregional rivalries and grievances, that there is little political capacity and enthusiasm left over for sustained and meaningful initiatives on the international front, especially as these relate to economic matters. By their essence federal structures are bound to ensure that regional matters are always high on the political agenda.

This speculation should not be taken as a rejection of the desirability of federation in Australia and Canada. Clearly, the federal structures of the two countries have provided enormous advantages in terms of administration and regional development in the broadest sense. There are some federal countries, such as West Germany, which have been extremely successful in developing their own strategies within the world economy. But as several chapters of this book point out, for their international posture, Australia and Canada prefer to rely on integration in the world economy even if, in practice and especially in relation to resolving internal interregional issues, this doctrine has been readily compromised. Many authors of this volume offer substantial reservations as to whether Australia and Canada should continue to rely on global integration to define their international roles, if historic expectations regarding income and employment are to be maintained. For economic geographers, an important lesson of this kind of comparative volume is the need to incorporate the role of the nation-state and of international relations in our thinking about development and location.

CONTRIBUTORS

John N.H. Britton is Professor of Geography, University of Toronto. His research is concerned with the impact of foreign ownership on Canadian economic development, with regional industrial and technological policy, and with the process of innovation.

James B. Cannon is Associate Professor in Geography at Queen's University. His research has focussed largely on the nature and effectiveness of industrial stimulation programs in Canada.

Bob Fagan is Senior Lecturer in Geography at Macquarie University where he teaches and researches economic, and particularly industrial restructuring. His publications focus on internationalization of capital, and the regional employment impacts of restructuring.

Katherine D. Gibson is a lecturer in the Department of Geography, University of Sydney. Her research interests include Marxian crisis theory, industrial restructuring, coal and migrant labour.

Thomas Gunton is an Associate Professor and Director of the Resource Management Program at Simon Fraser. His research principally examines public policy in the context of resource management

Roger Hayter is Professor and Chair, Department of Geography, Simon Fraser University. His research focusses on multinational corporations and regional development and the forest product industries.

John V. Langdale is Senior Lecturer in Geography at Macquarie University. His research interests are on the role of telecommunications in the international information economy especially with respect to such information service industries as banking and finance, computer services, media and business and financial information providers.

Richard LeHeron is Senior Lecturer in Geography, Massey University. His current research is emphasizing the local impacts of the internationalization of capital, especially with reference to New Zealand.

David Ley is Professor of Geography at the University of British Columbia, Vancouver. His research fields are urban and social geography, especially inner city issues. He is the author of several books including *A Social Ge-*

ography of the City (1983), and *The New Middle Class and the Restructuring of the Central City* (forthcoming).

Godfrey J.R. Linge is Professorial Fellow, Department of Human Geography, Australian National University. His current research interests are concerned with Australian industrial development, technology and regional development, and military spending and global industrialization.

T.G. McGee is Professor of Geography, and Director, Institute of Asian Research, University of British Columbia. He has carried out research on urbanization in Asia over the last 30 years and is currently working on labour force change in Malaysia.

John McKay is Senior Lecturer in Geography, Monash University. He is currently particularly interested in the spatial structure of labour force change in Australia.

Peter McLoughlin obtained his Ph.D. from Reading and is Assistant Director, Government Relations Branch, Office of Local Government (Australia). His research interests cover the effects of industrial restructuring on small businesses, local economic planning policies and intergovernmental relations in federal systems.

Caroline Mills received her B.A. from Oxford University and her Ph.D. in Geography at the University of British Columbia, Vancouver. Her research is concerned with the intersection of changing gender relationships and a post modern cultural sensibility in revitalizing inner city neighbourhoods.

Glen Norcliffe is an Associate Professor of Geography at York University, Toronto. His current research concerns rural labour markets in Ontario and the Massif Central.

David Rich is Senior Lecturer in Geography at Macquarie University. He teaches economic geography and urban and regional development. Recent publications include *The Industrial Geography of Australia,* London: Croom Helm, 1987. Current research focusses on industrial development in South Australia since 1930, and restructuring of the Australian sugar industry.

John Richards is an Assistant Professor in the Faculty of Business Administration, Simon Fraser University. A former member of the Saskatchewan legislature, his current research focusses on labour relations and resource policy.

Donna Smith Featherstone completed her M.A. at York University in 1986 and is currently a homemaker. Her interests are in regional economic disparities in Canada.

David F. Walker is Professor of Geography with cross appointment in the School of Urban and Regional Planning, University of Waterloo. His interests focus largely on regional and local area planning.

Iain Wallace is Associate Professor and Chairman, Department of Geography at Carleton University. His research interests focus on agribusiness, Canada's resource-based industries, geography and social theory.

Jim Walmsley is Associate Professor in the Department of Geography and Planning at the University of New England. His interests are in social, behavioural and political geography.

Peter D. Wilde is Senior Lecturer in Geography at the University of Tasmania and also Chairman of the Australian Study Group on Industrial Change, a study group of the Institute of Australian Studies. As well as publishing on Australia's role in the world economic system, his major research interest is examining change in Tasmania's economic activity and employment from radical perspectives.

REFERENCES

ADRIAN, C. and STIMSON, R. (1986) "Asian investments in Australian capital city property markets" *Environment and Planning A* 18, pp. 323-340.

AGLIETTA, M. (1979) *A Theory of Capitalist Accumulation.* London: New Left Books.

AITKEN, H.G.J. (1959) "Defensive expansionism: the state and economic growth in Canada" in EASTERBROOK, W.T. and WATKINS, M.H. (eds.) *Approaches to Canadian Economic History*. Toronto: McClelland and Stewart, pp. 183-221.

ALDRED, J. and WILKES, J. (eds) (1983) *A Fractured Federation?* Sydney: Allen and Unwin.

AMIN, A. and GODDARD, J. B. (1986) *Technological Change, Industrial Restructuring and Regional Development.* London: Allen and Unwin.

ANDERSON, K. and GARNAUT, R. (1987) *Australian Protectionism: Extent, Causes and Effects.* Sydney: Allen and Unwin.

ANDERSON, K., DRYSDALE, P., FINDLAY, L., PHILLIPS, P., SMITH., B. and TYERS, R. (1985) "Pacific economic growth and the prospects for Australian trade" *Pacific Economic Papers No. 122.* Canberra: Australia - Japan Research Centre.

ANDERSON, R. (1986) *Globe and Mail,* April 24 p. B2.

ANDREFF. W. (1984) "The international centralization of capital" *Capital and Class* 22, pp. 59-80.

ANGERS, F. A. (1972) "The multinational firm and the nation state: one view" in PAQUET, G. (ed.) *The Multinational Firm and the Nation State.* Don Mills: Collier, pp. 128-33.

ANONYMOUS (1986) "CHH partner in Chilean venture" *New Zealand Herald* May 10.

ANTONELLI, C. (1985) "The diffusion of an organizational innovation: international data telecommunications and multinational industrial firms" *International Journal of Industrial Organization* 3, pp.109-18.

APPENDICES TO THE JOURNAL OF THE HOUSE OF REPRESENTATIVES, (1926) *C3, Report of the New Zealand Forest Service.* Wellington: New Zealand, Government Printer.

ARMSTRONG, H. and TAYLOR, J. (1978) *Regional Economic Policy.* Deddington, Oxford: Philip Allen.

ARMSTRONG, W. R. and MCGEE T.G. (1985) *Theatres of Accumulation: Studies in Asian and Latin American Urbanization.* London: Methuen.

ARMSTRONG, W.R. (1978) "New Zealand: imperialism, class and uneven development" *Australia New Zealand Journal of Sociology* 14, pp. 297-303.

ARROW, K.J. (1962) "Economic welfare and the allocation of resources for invention" in *The Rate and Direction of Inventive Activity: Economic and Social Factors.* Princeton: National Bureau of Economic Research.

ATKINSON, M.M. and COLEMAN, W.D. (1989) *The State, Business and Industrial Change.* Toronto: University of Toronto Press.

AUCOIN, P. and BAKVIS, H.(1984) "Organizational differentiation and integration: the case of regional economic development policy" *Canadian Public Administration* 27, pp. 348-371.

AUCOIN, P. and BAKVIS, H.(1985) "Regional responsiveness and government organization: the case of regional economic development policy in Canada" in AUCOIN, P. (ed.) *Regional Responsiveness and the National Administrative State.* Research Study # 37, Royal Commission on the Economic Union and Development Prospects for Canada, Toronto: University of Toronto Press.

AUSTRALIA, COMMITTEE OF INQUIRY INTO TECHNOLOGICAL CHANGE IN AUSTRALIA (MYERS REPORT) (198O). *Report.* Canberra: Australian Government Publishing Service.

AUSTRALIA, COMMITTEE OF INQUIRY INTO THE AUSTRALIAN FINANCIAL SYSTEM (1981) *Final Report.* Canberra: Australian Government Publishing Service, 5 vols [The Campbell Report].

AUSTRALIA, COMMITTEE OF INQUIRY INTO THE AUSTRALIAN FINANCIAL SYSTEM (1984) *Report of the Review Group, December 1983* Canberra: Australian Government Publishing Service [The Martin Report].

AUSTRALIA, COMMITTEE ON INQUIRY INTO LABOUR MARKET PROGRAMMES (KIRKBY REPORT) (1985) *Report.* Canberra: Australian Government Publishing Service.

AUSTRALIA, DEPARTMENT OF IMMIGRATION, LOCAL GOVERNMENT AND ETHNIC AFFAIRS (1988) *Australia's Population Trends and Prospects.* Canberra: Australian Government Publishing Service.

AUSTRALIA, DEPARTMENT OF INDUSTRY, TECHNOLOGY AND COMMERCE (1984) *Information Kit: Motor Vehicle Policy.* Canberra: Australian Government Publishing Service.

AUSTRALIA, DEPARTMENT OF INDUSTRY, TECHNOLOGY AND COMMERCE (1985) *The Promotion of Indigenous R & D in Australia and the Effectiveness of the Industrial Research and Development Incentives Scheme,* Vol. 1, Canberra: Price Waterhouse.

AUSTRALIA, DEPARTMENT OF SCIENCE (1985) *National Technology Strategy: Revised Discussion Draft,* Canberra: Australian Government Publishing Service.

AUSTRALIA, DEPARTMENT OF SCIENCE AND TECHNOLOGY (1984) *National Technology Strategy: Discussion Draft.* Canberra: Department of Science and Technology.

AUSTRALIA, ECONOMIC PLANNING AND ADVISORY COUNCIL (1986) *International Trade Policy.* Canberra: Office of EPAC.

AUSTRALIA, PARLIAMENT (1977) *White Paper on Manufacturing Industry,* Canberra: Australian Government Publishing Service.

AUSTRALIA, REPORT TO THE PRIME MINISTER BY THE COMMITTEE TO ADVISE ON POLICIES (1987) *Report.* Canberra: Australian Government Publishing Service.

AUSTRALIA, SENATE STANDING COMMITTEE ON INDUSTRY AND TRADE (1985) *The Development of Closer Economic Relations between Australia and New Zealand; Third Report.* Canberra: Australian Government Publishing Service.

AUSTRALIA: PRICES SURVEILLANCE AUTHORITY (1986) *Inquiry Into the Chicken Meat Industry.* Canberra: Australian Government Publishing Service.

AUSTRALIAN COUNCIL OF TRADE UNIONS (1987) *Australia Reconstructed.* Canberra: Australian Government Publishing Service.

AUSTRALIAN ETHNIC AFFAIRS COUNCIL, (1977) *Australia as a Multicultural Society,* Canberra: Australian Government Publishing Service.

AUSTRALIAN POPULATION & IMMIGRATION COUNCIL/AUSTRALIAN ETHNIC AFFAIRS COUNCIL (1979) *Multiculturalism and Its Implications for Immigration Policy.* Canberra: Australian Government Publishing Service.

AUSTRALIAN SCIENCE AND TECHNOLOGY COUNCIL (ASTEC)

(1984) *Government Purchasing and Offsets Policies in Industrial Innovation,* Canberra: Australian Government Publishing Service.

BARNARD, J.T. (198O) *An Economic Analysis of the Canadian Taxation of the Integrated Mining Company.* Working Paper #18, Kingston, Ontario: Centre for Resource Studies, Queen's University.

BASSETT, K. and HAGGETT, P. (1971) "Towards short-term forecasting for cyclic behaviour in a regional system of cities" in CHISHOLM, M., FREY, A.W. and HAGGETT, P. (eds.) *Regional Forecasting.* London: Butterworth, pp. 389-413.

BEHRMANN, N. (1986) "Coal hit by South Africa's cut-throat prices" *Sydney Morning Herald.* 21st February 21, p. 19.

BELL, D. (1976) *The Coming of Post-Industrial Society.* New York: Basic Books.

BELL, D. (1980) *The Coming of Post-Industrial Society.* New York: Basic Books.

BELL, M.E. and LANDE, P.S. (1982) *Regional Dimensions of Industrial Policy.* Lexington: Lexington Books.

BELL, W. (1958) "Social choice, life styles and suburban residence" in DOBRINGER, W. (ed.) *The Suburban Community.* New York: Putnam, pp. 225-47.

BERCUSON, D.J. (ed.) (1977) *Canada and the Burden of Unity.* Toronto: Macmillan.

BERNARD, J.T. and PAYNE, R, (1987) "Natural resource rents and hydroelectric power" in GUNTON, T. and RICHARDS, J. (eds.) *Resource Rents and Public Policy.* Halifax: Institute for Research on Public Policy, pp. 59-84.

BERNSTEIN, J. I. (1985) "Research and development, patents and grant and tax policies in Canada" in MCFETRIDGE, D.G. (ed.) *Technological Change in Canadian Industry.* Toronto: University of Toronto Press.

BERRY, B. (1985) "Islands of renewal in seas of decay" in PETERSON, P. (ed.) *The New Urban Reality,* Washington, D.C.: The Brookings Institution, pp. 69-96.

BERTRAM, G.W. (1967) "Economic growth in Canadian industry 187O-1915: the staple model" in EASTERBROOK, W.T. and WATKINS, M.H. (eds.) *Approaches to Canadian Economic History,* Toronto: McClelland and Stewart, pp. 74-98.

BHP LTD (1986) *Reject the Offer for Your BHP Shares Made by Bell Resources.* Melbourne: Melbourne: BHP Ltd.

BLAINEY, G. (1984) *All for Australia* Melbourne: Methuen/Haynes.

BLUESTONE, B. and HARRISON, B. (1982) *The Deindustrialization of America.* New York: Basic Books.

BOADWAY, R. W. and FLATTERS, F. R. (1982) *Equalization in a Federal State.* Ottawa: Economic Council of Canada.

BORINS, S. F.(1986) *Investments in Failure: Five Government Enterprises That Cost the Canadian Taxpayer Billions.* Toronto: Methuen.

BRADBURY, J.H. (1985) "Regional and industrial restructuring processes in the new international division of labour" *Progress in Human Geography* 9, pp. 38-63.

BRADBURY, J.H. and ST-MARTIN, I. (1983) "Winding down in a Quebec mining town: a case study of Schefferville" *Canadian Geographer* 27, pp. 128-144.

BRADFIELD, M. (1977) "Living together: a review" *Canadian Public Policy* 3, pp. 504-509.

BRADFIELD, M.(1986) "Review essay: Macdonald Royal Commission Report" *Canadian Journal of Regional Science* 9, pp. 125-136.

BRANZI, A. (1984) *The Hot House: Italian New Wave Design.* Cambridge, Mass.: MIT Press.

BRECHLING, F.P.R. (1967), 'Trends and cycles in British regional unemployment" *Oxford Economic Papers* 19, pp. 1-21.

BRITTON, J.N.H. (1989a) "A policy perspective on incremental innovation in small and medium sized enterprises" *Entrepreneurship and Regional Development* 1, pp. 179-90.

BRITTON, J.N.H. (1989b) "Innovation policies for small firms" *Regional Studies* 23, pp. 167-73.

BRITTON, J.N.H. and GERTLER, M.S. (1986) "Locational perspectives on policies for innovation" in DERMER, J. (ed.) *Competitiveness Through Technology: What Business Needs from Government.* Lexington: Lexington Books, pp. 159-76.

BRITTON, J.N.H. and GILMOUR, J.M. (1978) *The Weakest Link A Technological Perspective on Canadian and Industrial Underdevelopment.* Science Council of Canada, Background Study 43. Ottawa: Supply and Services.

BROWETT, J. (1985) "Industrialization in the Global Periphery: The Significance of the Newly Industrializing Countries." *Discussion Paper* No. 10. The Flinders University of South Australia, Centre for Development Studies.

BRUCE-BRIGGS, B. (1979) *The New Class?* New Brunswick, N.J.: Transaction Books.

BRYM, J. and SACOUMAN R. (eds.) (1979) *Underdevelopment and Social Movements in Atlantic Canada.* Toronto: New Hogtown Press.

BUCH-HANSON, M. and NIELSON, B. (1977) "Marxist geography and the concept of territorial structure" *Antipode* 9, pp.1-12.

BULLOCK, M. (1983) *Academic Enterprise, Industrial Innovation, and the Development of High Technology Financing in the United States.* London: Brand Brothers & Co.

BUREAU OF AGRICULTURAL ECONOMICS (1984) *Economic Development in East and South-East Asia: Implications for Australian Agriculture in the 198Os.* Canberra: Australian Government Publishing Service.

BUREAU OF INDUSTRY ECONOMICS (1979) *The Australian Industrial Development - Some Aspects of Structural Change.* Research Report 2, Canberra: Australian Government Publishing Service.

BUREAU OF INDUSTRY ECONOMICS (1985) *The Regional Impact of Structural Change - An Assessment of Regional Policies: Australia, Research Report 18.* Canberra: Australian Government Publishing Service.

BUREAU OF INDUSTRY ECONOMICS (1986) *The Depreciation of the Australian Dollar: Its Impact on Importers and Manufacturers.* Canberra: AGPS.

BURLEY, K.H. (1960) "The overseas trade in NSW coal and the British shipping industry 1860-1914" *Economic Record* 36, pp.393-413.

BURNS, J., McINERNEY, J. and SWINBANK, A. (1983) *The Food Industry: Economics and Policies.* London: Heinemann.

BURNS, R.M. (1976) *Conflict and its Resolution in the Administration of Mineral Resources in Canada.* Kingston, Ont.: Centre for Resource Studies, Queen's University.

BUTLIN, N.G. (1959) "Colonial socialism in Australia" in AITKEN, M.G. (ed.) *The State and Economic Growth.* New York, Social Sciences Research Council, pp. 26-74.

BUTLIN, N.G. (1964) *Investment in Australian Economic Development.* Cambridge University Press.

BYRNES, M. (1986) "BHP amidst Kalimantan coal bonanza" *Australian Financial Review* April 2, p.8.

CAIN, J. (1983) *National Economic Summit Conference, Documents and Proceedings* (2). Canberra: Australian Government Publishing Service (3 vols).

CAMERON, D. and HOULE, F. (eds.) (1985) *Canada and the New International Division of Labour.* Ottawa: University of Ottawa Press.

CANADA (1966) *Report of the Royal Commission on Taxation.* Ottawa: Queen's Printer.

CANADA (1985) *Report of the Royal Commission on the Economic Union and Development Prospects for Canada.* Ottawa: Supply and Services Canada.

CANADA, (1985a) *Royal Commission on the Economic Union and Development Prospects for Canada.* Three volumes. Ottawa: Supply and Services.

CANADA, (1985b) *Intergovernmental Position Paper on the Principles and Framework for Regional Economic Development.* Ottawa: Supply and Services.

CANADA, DEPARTMENT OF COMMUNICATIONS (1982) *Telecommunications and Computer Services.* Working paper prepared for the Task Force on Trade in Services, Ottawa.

CANADA, MINISTERIAL TASK FORCE ON PROGRAM REVIEW (1986) *Services and Subsidies to Business: Giving with Both Hands.* Ottawa: Supply and Services.

CANADA, ENERGY MINES AND RESOURCES (1982) *Mineral Policy A Discussion Paper.* Ottawa: Supply and Services.

CANNON, J. B. (1984) "Explaining regional development in Atlantic Canada: a review essay" *Journal of Canadian Studies* 19, pp. 65-86.

CANNON, J.B. (1989) "Directions in regional policy" *The Canadian Geographer* 33, pp. 230-39.

CARDEW, R.V. and RICH, D.C. (1982) "Manufacturing and industrial property development in Sydney" in CARDEW, R.V., LANGDALE, J.V. and RICH, D.C. (eds.) *Why Cities Change: Urban Development and Economic Change.* Sydney: Allen and Unwin, pp. 115-34.

CARMICHAEL, E. A., DOBSON, W. and LIPSEY, R. G. (1986) "The Macdonald Report: signpost or shopping basket?" *Canadian Public Policy* 12, Supplement, pp. 40-50.

CARTER, R. (1983) "Policy for non-metropolitan regions: the case for a reassessment" *Australian Quarterly* Spring, pp. 319-333.

CARTER, S. (1986) "Sunset over the sunrise" *Australian Society* 3, pp. 8 - 11.

CASS, B. (1985) "Why women must screen tax changes" *Australian Society* 4, pp. 2O-3.

CASTELLS, M. (1983a) "Crisis, planning and the quality of life: managing the new historical relationships between space and society" *Environment and Planning D: Society and Space* 1, pp. 1-17.

CASTELLS, M. (1983b) *The City and the Grassroots*. London: Arnold.

CHAMBERS, E.J. and GORDON, D.F. (1966) "Primary products and economic growth: an empirical measurement" *Journal of Political Economy* 74, pp. 315-332.

CHAPMAN, K. and HUMPHRYS, G. (1987) *Technical Change and Industrial Policy*. London: Basil Blackwell.

CHESHIRE, P.C. (1973) *Regional Unemployment Differences in Great Britain*. Regional Papers, National Institute of Economic and Social Research, Cambridge University Press, Cambridge.

CHRISTOPHERSON, S. (1983) "The household and class formation: determinants of residential location in Ciudad Juarez" *Environment and Planning D Society and Space* 1, pp. 323-38.

CITY OF VANCOUVER (1984) "Jobs in Vancouver" *Quarterly Review* 11, pp. 5-7.

CITY OF VANCOUVER (1985) "Looking at downtown office space" *Quarterly Review* 12, pp. 17-19.

CLAIRMONTE, F.F. and CAVANAGH, J. H. (1984) "Transnational corporations and services: the final frontier" *Trade and Development An UNCTAD Review* No.5, pp.215-273.

CLARK, G. L. (1986) "The crisis of the Midwest auto industry" in SCOTT, A.J and STORPER M. (eds.) *Production, Work and Territory. The Geographical Anatomy of Industrial Capitalism*. Boston: Allen and Unwin, pp. 127-148.

CLARK, R. (1985) "Urban primacy: internal development or international relations? Nineteenth-century Australia" in TIMBERLAKE, M. (ed.) *Urbanization in the World-Economy*. Academic Press.

CLARKSON, S. (1985) *Canada and the Reagan Challenge*. Toronto: Lorimer.

CLAY, P. (1979) *Neighborhood Renewal: Middle Class Resettlement and Incumbent Upgrading in American Neighborhoods*, Lexington, Mass.: D.C. Heath.

CLEMENT, W. (1975) *The Canadian Corporate Elite*. Toronto: McClelland and Stewart.

CLEMENT, W. (1977) *Continental Corporate Power*. Toronto: McClelland and Stewart.

CLOW, M. (1984) "Politics and uneven capitalist development: the Maritime challenge to the study of Canadian political economy" *Studies in Political Economy* 14, Summer, pp. 117-40.

COAKLEY, J. (1984) "The internationalization of bank capital" *Capital and Class* 23, pp. 107-120.

COHEN, J. and KRUSHINSKY, M. (1976) "Capturing resource rents for the public land owner: the case for a crown coporation" *Canadian Public Policy* 2, pp. 411-423.

COHEN, R.B. (1981) "The new international division of labor, multinational corporations and the urban hierarchy" in DEAR, M. and A.J. SCOTT (eds.) *Urbanization and Urban Planning in Capitalist Societies*. New York: Methuen, pp. 287-315.

COMMITTEE FOR THE ECONOMIC DEVELOPMENT OF AUSTRALIA (1986) *Exploring Opportunities in the Pacific Basin*. Melbourne: Committee for the Economic Development of Australia.

COMMITTEE ON INQUIRY INTO TECHNOLOGICAL CHANGE IN AUSTRALIA (198O) *Technological Change in Australia*. Canberra: Australian Government Publishing Service.

CONLON, R.M. (1985) *Distance and Duties*. Ottawa: Carleton University Press.

CONNELL, R.W. (1977) *Ruling Class Culture: Studies of Conflict, Power and Hegemony in Australian Life*. Cambridge: Cambridge University Press.

COOK, L. and DIXON. P. (1982) "Structural change and employment prospects for migrants in the Australian work force" *Australian Economic Papers* 21, pp. 69-94.

COOMBS, H.C. (1981) *Trial Balance.* Melbourne: Macmillan.

COPITHORNE, L. (1979) *Natural Resources and Regional Disparities.* Ottawa: Economic Council of Canada.

CORDELL, A.J. and GILMOUR, J.M. (1986) *The Role and Function of Government Laboratories and the Transfer of Technology to the Manufacturing Sector* Ottawa: Science Council of Canada, Background Study 35.

CORDEY-HAYES, M. and GLEAVE, D. (1974) "Migration movements and the differential growth of city regions in England and Wales" *Papers of the Regional Science Association* 33, pp. 99-123.

COURCHENE T. J. (1978) "Avenues of Adjustment: the Transfer System and Regional Disparities." in WALKER, M (ed.) *Confederation at the Crossroads.* Vancouver: Fraser Institute, pp. 145-185.

COURCHENE T.J. (1983) "Canada's New Equalization Program: Description and Evaluation" *Canadian Public Policy* 9, pp. 458-75.

COURCHENE T.J. (1984) *Equalization Payments: Past, Present and Future.* Toronto: Ontario Economic Council.

COURCHENE, T. J. and MELVIN, J. R. (1986) Canadian regional policy: lessons from the past and prospects for the future *Canadian Journal of Regional Science 9*, pp. 49-68.

CRISP, J. (1986) "Cable and wireless prepares to take on the world" *Financial Times* 23 June, p. 5.

CROUGH, G. (1988) 'Transnational corporations and foreign direct investment in Australia', a paper delivered to APDC-RIAP Conference, University of Sydney, July 14.

CROUGH, G. AND WHEELWRIGHT, T. (1982) *Australia: A Client State,* Ringwood Victoria: Penguin Books.

DALES, J.H., MCMANUS J.C. and WATKINS, M.H. (1967) "Primary products and economic growth: a comment" *Journal of Political Economy* 75, pp. 876-88O.

DALY, M.T. (1968) "The lower Hunter Valley urban complex and the dispersed city hypothesis" *The Australian Geographer* 19, pp.472-482.

DALY, M.T. (1982) *Sydney Boom Sydney Bust.* Sydney: Allen and Unwin.

DALY, M.T. (1984) "The revolution in international capital markets: urban growth and Australian cities" *Environment and Planning A* 16, pp. 1003-1020.

DALY, M.T. (1987) "Rationalization of international banking and the implications for the Pacific Rim." Paper presented to the *Conference on Transnational Capital and Urbanization on the Pacific Rim,* Centre for Pacific Rim Studies, University of California, Los Angeles, March 26-28.

DALY, M.T. and LOGAN, M.I. (1986) "The international financial system and national economic development patterns " in DRAKAKIS-SMITH, D. (ed) *Urbanization in the Developing World.* London: Croom Helm, pp. 37-62.

DALY, M.T. and LOGAN, M.I. (1989) *The Brittel Rim: Finance, Business and the Pacific Region* (Penguin, Ringwood).

DANIELS, P. (1985) *A Geography of Unemployment in Vancouver C.M.A.,* Unpublished M.A. Thesis, University of British Columbia, Department of Geography.

DANIELS, P.W. (1985) *Service Industries: A Geographical Appraisal,* London: Methuen.

DARGAVEL, J. HOBLEY, M. and KENGEN, S. (1985) "Forestry of development and underdevelopment of forestry" in DARGAVEL, J. and SIMPSON, G. (eds.) *Forestry: Success or Failure in Developing Countries?* Centre for Resource and Environmental Studies, WP 1985/20, Department of Forestry. Canberra: Australian National University.

DAVIES, A. (1986) "Singapore office for Freehills" *Australian Financial Review* 3 July, p.27.

DE GRANDPRE REPORT (1989) *Adjusting to Win* Ottawa: Report of the Advisory Council or Adjustment.

DEAR, M. and SCOTT, A.J., (eds.) (1980) *Urbanization and Urban Planning in Capitalist Society.* London: Methuen.

DELWORTH, W.T. (1982) "Canada and the Pacific Rim: a political perspective" in QUO, F.G. (ed.) U Politics of the Pacific Rim: Perspectives on the 1980s. Burnaby, Simon Fraser University, pp. 3-10.

DEPARTMENT OF IMMIGRATION AND ETHNIC AFFAIRS (1984) *Statement Prepared for the International Conference on Population, Mexico 1984,* Canberra: AGPS.

DEPARTMENT OF LOCAL GOVERNMENT, AND ADMINISTRATIVE SERVICES, (1987) *Regional Variations in Manufacturing Industry Structure and Performance,* Australian Regional Developments 4.5, Canberra: AGPS.

DOERN, G. B. (1986) "The Tories, free trade and industrial adjustment policy: expanding the state now to reduce the state later?" in PRINCE, M.J. (ed.) *How Ottawa Spends 1986-87: Tracking the Tories.* Toronto: Methuen, pp. 61-92.

DONALDSON, M. (1985) "The Case For Wollongong" *Urban Policy and Research* 3, pp.38-40.

DONNER, A. (1986) "Concept of full employment a thing of the past" *Sunday Star* Toronto, July 20, F4.

DOUGLASS, M. (1986) "Structural change in the Pacific Rim: perspective on Japan." Paper presented to the Conference on *Industrial Transformation and Challenges in Australia and Canada International Perspectives* Simon Fraser University, Vancouver, August 18-22.

DOUGLASS, M. (1987) "Transnational capital and urbanization in Japan." Paper presented to the Conference on *Transnational Capital and Urbanization on the Pacific Rim,* Centre for Pacific Rim Studies, University of California, Los Angeles, March 26-28.

DOYLE, D. (1986) *Technology Venturing in Canada.* Ottawa: Ministry of State for Science and Technology.

DRACHE, D. (1978) "Rediscovering Canadian political economy" in CLEMENT, W. and DRACHE, D. (eds) *A Practical Guide to Canadian Political Economy.* Toronto: Lorimer, pp. 1-53.

DRACHE, D. and CLEMENT, W. (.1986) *Practical Guide To Canadian Political Economy.* Toronto: Lorimer

DRUCKER, P.F. (1986) "The changed world economy" *Foreign Affairs* 64 pp. 768-91.

DUNCAN, T. and FOGARTY, J. (1984) *Australia and Argentina: On Parallel Paths.* Melbourne: Melbourne University Press.

DUTHIE, S. (1986), "Simex looks to futures expansion" *Australian Financial Review* July 14, p. 3.

ECONOMIC COUNCIL OF CANADA (1975) *Looking Outward.* Ottawa. Economic Council of Canada.

ECONOMIC COUNCIL OF CANADA (1977) *Living Together: A Study of Regional Disparities.* Ottawa: Economic Council of Canada.

ECONOMIC COUNCIL OF CANADA (1982) *Financing Confederation.* Ottawa: Supply and Services.

ECONOMIC PLANNING ADVISORY COUNCIL (1986) *Regional Impact of Industry Assistance.* Canberra: EPAC Council Paper No. 2O.

EDWARDS, C.T. (1982) "The impact of economic change in Asia on Australia" in WEBB, L.R. and ALLAN, R.H. (eds.) *Industrial Economics: Australian Studies.* Sydney: Allen and Unwin, pp. 443-61.

EDWARDS, M., HARPER, T., and HARRISON, M. (1985) "Child support: public or private duty" *Australian Society* 4, pp. 18-22.

ELDERS IXL LTD (1986) *Annual Report 1986*, Adelaide: Elders IXL Ltd.

ELLIGOTT, R. and ZACHARIAS, J. (1973) *Fairview Slopes*, Vancouver: Department of Planning and Civic Development, City of Vancouver.

ELLIOTT, D.AS. (1982) "Review of thinning practice in New Zealand 1974 to 1981" *New Zealand Journal of Forestry Science* 2, pp. 127-139.

ELLIS, L.M. (1920) *Report on Forest Conditions in New Zealand, and a Case for a National Forest Policy.* Appendices to the Journal of the House of Representatives, C3A.

EMERY, F.E. and TRIST, E.L. (1973) *Towards a Social Ecology*, New York: Plenum.

EMERY, F.E., EMERY, M., CALDWELL, G. and CROMBIE, A. (1974) *Futures We're In. Canberra*: CCAE.

ESPIE REPORT (1983) *Developing High Technology Enterprises for Australia.* Parkville, A.C.T.: Australian Academy of Technological Sciences.

FAGAN, R.H. (1984) "Corporate structure and regional uneven development in Australia: the case of BHP Ltd," in Taylor, M. ed. *The Geography of Australian Corporate Power* (Croom Helm, Australia), pp. 91-123

FAGAN, R.H. (1987) "Australia on the periphery: geographical perspectives on economic reorganization" in Jeans, D.N. (ed.) *Australia - A Geography: Vol 2: Space and Society.* Sydney: Sydney University Press, pp. 400-425.

FAGAN, R.H. (1988) "Australia in the global economy" a paper delivered to the International Geographical Union, Commission on Industrial Change, Rutherglen Conference, Tasmania, August 14-20

FAGAN, R.H. (1990) "Elders IXL Ltd: finance capital and the geography of corporate restructuring," *Environment and planning A* (forthcoming).

FAGAN, R.H., MCKAY, J.H. and LINGE, G.J.R. (1981) "Structural change in the international and national context" in LINGE, G.J.R. and MCKAY, J.H. (eds.) *Structural Change in Australia. Some Spatial and Organizational Responses.* Department of Human Geography. Publication H.G./15, Canberra, ANU, pp. 1-52.

FISHER, C. (1987) *Coal and the State.* Sydney: Methuen.

FLETCHER CHALLENGE LIMITED (1981-86) *Annual Reports and Accounts.*

FLETCHER HOLDINGS (1978) "Fletcher Backgrounder" *Fletcher Industries Annual Report.*

FLETCHER, H.A. (1982) "The Role of the Private Sector in Resource Development" *Proceedings 11th New Zealand Geography Conference,* pp. 5-7.

FOGELBERG, G. (1984) "Acquisitions and their impact upon strategy and performanc: a preliminary investigation of the New Zealand experience". Paper presented to the *Strategic Management Conference,* Philadelphia, USA, October 11.

FORBES, E. R. (1979) *The Maritime Rights Movement - 1912-27: A Study in Canadian Regionalism.* Montreal: McGill-Queen's Press.

FRANKLIN, S.H. (1978) *Trade, Growth and Anxiety.* Sydney: Methuen.

FREEMAN, C. (1986) "The role of technical change in national economic development" in AMIN, A. and GODDARD, J.B. (eds.) *Technological Change, Industrial Restructuring and Regional Development,* London: Allen and Unwin, pp. 100-113.

FRENCH, R. D. (1980) *How Ottawa Decides.* Ottawa: Canadian Institute for Economic Policy.

FRIEDMAN, J. (1972) "A general theory of polarized development" in HANSEN, M.N. (ed) *Growth Centres in Regional Economic Development.* New York: Free Press, pp. 82-107.

FRIEDMANN, J. (1986) "The world city hypothesis" *Development and Change* 17, pp. 69-83.

FRIEDMANN, J. and WEAVER, C. (1979) *Territory and Function: The Revolution of Regional Planning.* London: Edward Arnold.

FRIEDMANN, J. and WOLFF, G. (1982) "World city formation: An agenda for research and action" *International Journal of Urban and Regional Research* 6, pp. 309-43.

FRIZZELL, A., PAMMETT, J.H. and WESTELL, A., (1989) *The Canadian General Election of 1988.* Ottawa: Carleton University Press.

FROBEL, F., HEINRICHS, J. and KREYE, O. (1980) *The New International Division of Labour. Structural Unemployment in Industrialized Countries and Industrialization in Developing Countries.* London: Cambridge University Press.

FUJII, G. (1981) *The Revitalization of the Inner City: A Case Study of the Fairview Slopes Neighborhood, Vancouver, B.C.*, Unpublished M.A. Thesis, University of British Columbia, Department of Geography.

GALE, D. (1984) *Neighborhood Revitalization and the Post-Industrial City.* Lexington, Mass.: D.C. Heath.

GALOIS, R.M. and MABIN, A. (1981) "Canada, the United States and the world-system: the metropolis-hinterland paradox" in McCann, L.D. (ed) *A Geography of Canada: Heartland and Hinterland.* Scarborough: Prentice-Hall Canada, pp. 37-62.

GANNICOTT, G.K. (1982) "Research and development incentives" in PARRY, T.G. (ed.) *Australian Industrial Policy.* London: Longman.

GARNAUT, R. (1989) *Australia and the Northeast Asian Ascendancy* (Report to the Prime Minister for Minister for Foreign Affairs and Trade) Australian Government Publishing Service, Canberra.

GARNAUT, R. and ANDERSON, K. (198O) "ASEAN export specialization and the evolution of comparative advantage in the Western Pacific region" in GARNAUT, R. (ed.) *ASEAN in a Changing Pacific and World Economy.* Canberra: Australian National University Press, pp. 374-412.

GARNER, H. (1976) *The Intruders.* Toronto: McGraw-Hill Ryerson.

GEE, M. (1978) "The avenue recultured" *Vancouver Province* May 25, p.15.

GIBSON, K.D. (1984) "Industrial organization and coal production in Australia, 186O-1982: an historical materialist analysis" *Australian Geographical Studies* 22, pp. 221-242.

GIBSON, K.D. (1990) "Australian coal in the global context: a paradox of efficiency and crisis" *Environment and Planning A* 22, (forthcoming).

GIBSON, K.D. and HORVATH, R.J. (1983) "Global capital and the restructuring crisis in Australian manufacturing" *Economic Geography* 59, pp. 178-194.

GIBSON, K.D. and Horvath, R.J. (1983) "Aspects of a theory of transition within the capitalist mode of production" *Environment and Planning D, Society and Space* 1, pp.121-138.

GILMOUR, J.M. (1982). "The industrial policy debate in a resource hinterland" *Search* 13, pp. 7-8.

GLOBE AND MAIL (1987) "Central Canadians emerge big winners" June 5, p. A12.

GLOBE AND MAIL (1988) "Atlantic provinces await aid" January 28, p. A4.

GODDARD, J. (1980) "Technological forecasting in a spatial context" *Futures* 12, pp. 90-105.

GOLLAN, R. (1960) *The Coal Miners of New South Wales: A History of the Union.* Melbourne: Melbourne University Press.

GONICK, C. (1986) *The Great Economic Debate: Failed Economics and A Future for Canada,* Toronto: Lorimer.

GOULDNER, A. (1979) *The Future of Intellectuals and the Rise of the New Class.* New York: Seabury.

GRANT, B. (1983) *The Australian Dilemma: A New Kind of Western Society.* Sydney: McDonald Futura.

GRANT, P. (1983) "Technological sovereignty: forgotten factor in the 'Hi-Tech' razzamatazz" *Prometheus* 1, pp.239-270.

GRAY REPORT (1972) *Foreign Direct Investment in Canada.* Ottawa: Government of Canada.

GREGORY, R.G. (1985) "Industry protection and adjustment: the Australian experience" *Prometheus* 3, pp. 25-50.

GROENEWEGEN, P.D. (1983) "The political economy of federalism, 1901-81" in Head, B.W.(ed.) *State and Economy in Australia.* Melbourne: Oxford University Press, pp. 169-195.

GRUFT, A. (1983) "Vancouver architecture: the last fifteen years" in *Vancouver: Art and Artists 1931-1983.* Vancouver, B.C. Vancouver Art Gallery, pp. 318-31.

GUNTON, T. (1982) *Resources, Regional Development and Public Policy: A Case Study of British Columbia.* Paper #8. Ottawa: Centre for Policy Alternatives.

GUNTON, T. (1987) "Manitoba's nickel industry: the paradox of a low cost producer" in GUNTON, T. and RICHARDS, J. (eds.) *Resource Rents and Public Policy.* Halifax: Institue for Research on Public Policy.

GUNTON, T. and RICHARDS, J. (eds), (1987) *Resource Rents and Public Policy.* Halifax: Institute for Research on Public Policy.

HARRIS, R. C. (1982) "Regionalism and the Canadian archipelago" in McCann, L. D. (ed.), *Heartland and Hinterland: A Geography of Canada.* Toronto: Prentice-Hall Canada, pp. 458-84.

HAVAS, T. (1986) "Regional Approaches - A Commonwealth Perspective" paper presented to the *National Conference on Resource Development and Local Government: Policies for Growth, Decline and Diversity,* Canberra, Australia.

HAYASHI, K. (1980) "The role of the multinational enterprise in Canada-Japan relations" in HAY, K.A.J. (ed.) (*Canadian Perspectives on Economic Relations with Japan*). Montreal: Institute for Research on Public Policy, pp. 163-87.

HAYTER, R. (1983) *Canada's Pacific Basin Trade and Its Implications for the Exports of Manufacturers in B.C. and Alberta. Preliminary Perspectives.* Project: Canada and the Changing Economy of the Pacific Basin, Institute of Asian Research, Working Paper No. 11. Vancouver: University of British Columbia.

HAYTER, R. (1986) "Export performance and export potentials: western Canadian exports of manufactured end products" *Canadian Geographer* 30, pp. 26-39.

HEAD, B.W. (1984) "Australian resource development and the national fragmentation thesis" *Australian and New Zealand Journal of Sociology* 20, pp. 306-31.

HEADLAM, F. (1985) "Working wives" *Australian Society* 4, pp. 5-7.

HEALY, B. (1982) *A Hundred Million Trees. The Story of NZ Forest Products Ltd.* Sydney: Hodder and Stoughton.

HENDERSON, R. and HOUGH, D. (1984) "Sydney's poor get squeezed" *Australian Society* 3, pp. 6-8.

HIGGINS, B. (1986) *The Rise and Fall? of Montreal.* Moncton: Canadian Institute for Research on Regional Development.

HIGGINS, B.H. (1981) "Economic development and regional disparities: a comparative study of four federations" in MATTHEWS, R.C.(ed) *Regional Disparities and Economic Development.* Canberra: Centre for Research on Federal Financial Relations, pp. 21-8O.

HIGGOTT, R. (1987) "The dilemmas of interdependence: Australia and the international division of labour in the Asia-Pacific region" in CAPORASO, J.A. (ed.) *A Changing International Division of Labour.* London: France Pinter.

HILL, S. and JOHNSTON, R. (eds.) (1983) *Future tense? Technology in Australia.* Brisbane: UQP.

HINCE, K. (1982) *Conflict and Coal: A Case Study of Industrial Relations in the Open-cut Coal Mining Industry of Central Queensland.* St Lucia: University of Queensland Press.

HIRST, J. and TAYLOR, M.J. (1985) "Internationalization of Australian banking: further moves by the ANZ" *Australian Geographer* 16, pp. 291-95.

HIRST, J., TAYLOR, M.J. and THRIFT, N.J. (1982) "The geographical pattern of the Australian trading banks, overseas representation" in TAYLOR, M.J. and THRIFT, N.J. (eds.) *The Geography of Multinationals.* London: Croom Helm, pp. 117-35.

HO, S.P.S. and HUENEMANN, R.W. (1093) *China's Open Door Policy.* Canada and the Changing Economy of the Pacific Basin, Working Paper No. 4, Institute of Asian Research, University of British Columbia.

HOLDSWORTH, D. (1983) "Appropriating the past: heritage designation and inner city revitalization" Paper presented to the *Annual Conference of the Canadian Association of Geographers,* Winnipeg.

HOLMES, J. (1986) "The organization and locational structure of production sub-contracting" in SCOTT, A.J. and STORPER, M. (eds.) *Production, Work and Territory: the Geographical Anatomy of Industrial Capitalism.* Boston: Allen and Unwin, pp. 8O-106.

HOWARD, C. (1977) "The constitution as a legal document" in ENCEL, S., HORNE, D. and THOMPSON, E.(eds) *Change the Rules! Towards a Democratic Constitution.* Ringwood: Penguin, pp. 21-42.

HUGO, G.J. and SMAILES, P.J. (1985) "Urban-rural migration in Australia: a process study of the turnaround" *Journal of Rural Studies* 1, pp. 11-3O.

HUTTON, T. and LEY, D. (1987) "Location, linkages and labour: the downtown complex of corporate activities in a medium size city" *Economic Geography* 63, pp. 126-41.

IMS (INSTITUTE OF MANPOWER STUDIES) (1986) *UK Occupations and Employment Trends to 1990.* London: Butterworths.

INDUSTRIES ASSISTANCE COMMISSION, (1981) *The Regional Implications of Economic Change.* Discussion Paper No. 3, Canberra: AGPS.

INNIS, H.A. (1972) *Empire and Communications.* Toronto: University of Toronto Press.

INTERREGIONAL PLANNING GROUP (1983) *Central North Island Planning Study Findings.* Wellington: Ministry of Works and Development.

JAGER, M. (1986) "Class Definition and the Esthetics of Gentrification: Victoriana in Melbourne" in SMITH, N. and WILLIAMS, P. (eds.) *Gentrification of the City,* London: Allen and Unwin, pp. 78-91.

Migration 1966-1971 to 1971-1976. Unpublished Ph.D. thesis, Flinders University, Adelaide.

JARVIE, W.K. and BROWETT, J.G. (1980) "Recent changes in migration patterns in Australia" *Australian Geographical Studies* 18, pp. 135-145.

JEANS, D.N. and SPEARITT, P. (1980) *The Open Air Museum* Sydney: Allen and Unwin.

JENKINS, M. (1983) *The Challenge of Diversity: Industrial Policy in the Canadian Federation.* Ottawa: Science Council of Canada.

JENKINS, R. (1984) "Divisions over the international division of labour" *Capital and Class* 22, pp. 28-58.

JENSEN, R.C. and WEST, G.R. (1985) *A Preliminary Survey of Australian Regional Input-Output Models.* Canberra: DOLGAS.

JOHNSTON, R. and RUTNAM, R. (1981) "The effects of technological change on employment in the Wollongong region in the 1980s" *Report in the Science, Technology and Public Policy Series,* Department of History and Philosophy of Science, University of Wollongong.

JOINT COAL BOARD, (1978-79 to 1984-85) *Black Coal in Australia: A Statistical Yearbook.* Sydney: Joint Coal Board.

JONES, B. (1983) *Sleepers, Awake.* Melbourne: Oxford University Press.

JOSEPH, R.A. (1984) "Recent trends in Australian government policies for technological innovation" *Prometheus* 2, pp. 93-111.

JOSEPH, R.A. (1989) "Technology parks and their contribution to the development of technology oriented complexes in Australia" *Environment and Planning C* 7, pp. 173-92.

JOSEPH, R.A. and JOHNSTON, R. (1985) "Market failure and government Support for Science and Technology: Economic Theory versus Political Practice" *Prometheus* 3, pp.138-155.

KASPER, W. (1984) *Capital Xenophobia: Australia's Controls of Foreign Investment.* St. Leonards, N.S.W.: Centre for Independent Studies.

KASPER, W., BLANDY, R., FREEBAIRN, J., HOCKING, D., O'NEILL, R. (1980) *Australia at the Crossroads.* Sydney: Harcourt, Brace Jovanovich.

KELLERMAN, A. (1984) 'Telecommunications and the geography of metropolitan areas" *Progress in Human Geography* 8, pp. 222-46.

KELLY, E.R.R. and LECRAW, D. (1985) "Trading houses to spur Canadian exports" *Business Quarterly* Spring, pp. 36-42.

KENNEDY, R. (1986) "Potter links with international firms" *Sydney Morning Herald* 4, March, p. 29.

KIERANS, E. (1973) *Report on Natural Resources Policy in Manitoba.* Winnipeg: Government of Manitoba.

KILJUNEN, M. J. (1980) "Regional disparities and policy in the EEC" in SEERS, D. and VAITSOS, C. (eds.) *Integration and Unequal Development: The Experience of the EEC.* New York: St. Martins Press, pp. 199-222.

KING, L.J. and CLARK, G.L. (1978) "Regional unemployment patterns and the spatial dimensions of macro-economic policy: the Canadian experience 1966-1975" *Regional Studies* 12, pp. 283-296.

KIRBY REPORT (1985) *Australia, Committee of Inquiry into Labour Market Programs.* Canberra: AGPS.

KLEIN, S. (1987) "Export promotion: the trading house option revisited" *Canadian Public Policy* 12, pp. 284-93.

KORPORAAL, G. (1985) "Aussie brokers battle for US business" *Australian Financial Review* December 6, p.16.

LANGDALE, J.V. (1982a) 'Telecommunications in Sydney: towards an information economy" in CARDEW, R., LANGDALE, J.N., and RICH, D., (eds.) *Why Cities Change: Urban Development and Economic Change in Sydney.* Sydney: Allen and Unwin, pp. 77-94.

LANGDALE, J.V. (1982b) "Competition in telecommunications" *Telecommunications Policy* 6, pp. 283-99.

LANGDALE, J.V. (1984a) "Information services in Australia and Singapore" *ASEAN-Australian Economic Papers*, No. 16, Canberra and Kuala Lumpur: ASEAN-Australia Joint Research Project.

LANGDALE, J.V. (1984b) "Computerization in Singapore and Australia" *The Information Society* 3, pp. 131-53.

LANGDALE, J.V. (1985a) "Electronic funds transfer and the internationalization of the banking and finance industry" *Geoforum* 16, pp. 1-13.

LANGDALE, J.V. (1985b) *Transborder Data Flow and International Trade in Electronic Information Services: An Australian Perspective,* Canberra: Australian Government Publishing Service.

LANGDALE, J.V. (1987) "Telecommunications and electronic information services in Australia" in BROTCHIE, J.F., HALL, P., and NEWTON, P.W. (eds.) *Spatial Impact of Technological Change,* London: Croom Helm, pp. 89-103.

LANGDON, F. (1983) *The Politics of Canadian-Japanese Economic Relations, 1952-1983*, Vancouver: UBC Press.

LASH, S. and URRY, J. (1987) *The End of Organized Capitalism.* Cambridge: Polity Press.

LAWRENCE, G. (1987) *Capitalism and the Countryside.* Sydney: Pluto Press.

LAWRENCE, M. (1986) "The Riverland Development Council" paper presented to the *National Conference on Industry Policy, Employment and Regional Development*, Newcastle.

LEHERON, R.B. (1980) "The diversified corporation and development policy: New Zealand's experience" *Regional Studies* 14, pp. 201-217.

LEHERON, R.B. (1987a) "Changing private-state relations during an era of exotic afforestation, 1960-1965" *Proceedings of the13th New Zealand Geography Conference.*

LEHERON, R.B. (1987b) "Rethinking regional development" in JOHNSTONE, W.B. AND HOLLAND, P. (eds.) *Southern Approaches: Geography in New Zealand.* Christchurch: New Zealand Geographical Society.

LEHERON, R.B. and ROCHE, M.M. (1985) "Expanding exotic forestry and the extension of a competing use for rural land in New Zealand" *Journal of Rural Studies* 1, pp. 211-229.

LEROY, D.J. and DUFOUR, P. (1983) *Partners in Industrial Strategy.* Ottawa: Science Council of Canada, Background Study No. 51.

LEVER, W.F. (1980) 'The operation of local labour markets in Great Britain " *Papers of the Regional Science Association* 44, pp. 37-55.

LEWIS, F. (1975) "The Canadian wheat boom and per capita income: new estimates" *Journal of Political Economy* 83, pp. 1249-1257.

LEY, D. (1983) *A Social Geography of the City.* New York: Harper and Row.

LEY, D. (1985a) "Work-residence relations for head office employees in an inflating housing market" *Urban Studies* 22, pp. 21-38.

LEY, D. (1985b) *Gentrification in Canadian Inner Cities: Patterns, Analysis, Impacts and Policy*, Ottawa: Canada Mortgage and Housing Corporation.

LEY, D. (1986) "Alternative explanations for inner city gentrification: a Canadian assessment" *Annals, Association of American Geographers* 76, pp. 521-35.

LEY, D. (1988) "Social upgrading in six Canadian inner cities" *The Canadian Geographer* 32, pp. 31-45.

LEY, D. and HUTTON, T. (1987) "Vancouver's corporate complex and producer services sector: linkages and divergence within a provincial staples economy" *Regional Studies* 21, pp. 413-24.

LINGE, G.J.R. (1967) "Governments and the location of secondary industry in Australia" *Economic Geography* 43, pp. 43-63.

LINGE, G.J.R. (1979) "Australian manufacturing in recession: a review of the spatial implications" *Environment and Planning A* 11, pp. 1405-1430.

LINGE, G.J.R. (1979) *Industrial Awakening: A Geography of Australian Manufacturing 1788 to 1890*. Canberra: ANU Press.

LINGE, G.J.R. (1988) "Australian space and global space" in Heathcote, R.L and Mabbutt, J.A. (eds.) *Land Water and People: Geographical Essays in Australian Resource Management*. Allen and Unwin, Sydney, pp. 239-260

LIPIETZ, A. (1985) "Les transformations dans la division internationale du travail", in CAMERON, D. and HOULE, F. (eds.) *Canada and the New International Division of Labour*. Ottawa: University of Ottawa Press, pp. 27-55.

LIPTON, S.G. (1977) "Evidence of central city revival" *Journal, American Institute of Planners* 43, pp. 136-47.

LITHWICK N. H. (1982a) "Canadian regional policy: undisciplined experimentation" *Canadian Journal of Regional Science* 5, pp. 275-282.

LITHWICK, N. H. (1982b) "Regional policy: the embodiment of contradictions" in DOERN, G.B. (ed.) *How Ottawa Spends your Tax Dollars*. Toronto: Lorimer, pp. 131-46.

LITHWICK, N. H. (1982c) "Regional policy: a matter of perspective" *Canadian Journal of Regional Science* 5, pp. 353-363.

LITHWICK, N. H. (1986) "Federal government regional economic development policies: an evaluative survey" in NORRIE, K., research co-ordinator, *Disparities and Interregional Adjustment*. Research Study #64, Royal Commission on the Economic Union and Development Prospects for Canada, Toronto Press, pp. 109-57.

LITUAK, I.A. and MAULE, C.J. (1981) *The Canadian Multinationals*. Toronto: Buterworth.

LLOYD, C. and TROY, P. (1984) "Duck Creek revisited? The case for national urban and regional policies" in HALLIGAN, J. and PARIS, C. (eds) *Australian Urban Politics: Critical Perspectives*. Melbourne: Longman Cheshire, pp. 45-57.

LLOYD, T. (1966) "A water resource policy for Canada" *Canadian Geographic Journal* 72, pp. 2-17.

LOCAL GOVERNMENT MINISTERS' CONFERENCE, (1987) *The Role of Local Government in Economic Development - Discussion Paper.* Canberra: AGPS.

LOGAN, A. (1979) "Recent directions of regional policy in Australia" *Regional Studies* 11, pp. 153-160.

LOGAN, M.I. (no date) "Australia's Place in the World System".

LOGAN, M.I. and MCKAY, J. (198O) "Regional development policy and population redistribution" in BURNLEY, I.H., PRYOR, R. and ROWLAND, D.T. (eds.) *Mobility and Community Change in Australia.* Brisbane: University of Queensland Press, pp. 232-249.

LOGAN, M.I., WHITELAW, J.S., and MCKAY, J. (1981) *Urbanization: the Australian Experience.* Melbourne: Shillington House.

LORIMER, J. (1986) "Ottawa is quietly shutting down regional development" *Atlantic Insight* 8, p. 3.

LOUGHEED, A.L. (1988) *Australia and the World Economy.* Ringwood Penguin.

LUM, S. (1984) *Residential Redevelopment in the Inner City of Vancouver: a Case Study of Fairview Slopes,* Unpublished MPl. Thesis, Queen's University, School of Urban and Regional Planning.

MACDONALD, R. (1987) "Picking up the pieces: Australia's future in the balance" *Business Review Weekly* 27 November, pp. 22-5.

MACDONALD, S (1983) "High technology policy and the Silicon Valley Model: an Australian perspective" *Prometheus* 1, pp. 330-349.

MACINTOSH, W.A. (1964) *The Economic Background of Dominion-Provincial Relations,* Appendix III of the Royal Commission on Dominion-Provincial Relations. Toronto: McClelland and Stewart. (First published 1939).

MACLEAN'S (1986a) "*A Troubled Province*" August 11, p. 15.

MACLEAN'S (1986b) "*Quebec's New Entrepreneurs*" August 4, pp. 24-25.

MACLEOD, G. (1987) *New Age Business: Community Corporations That Work.* Toronto: Lorimer.

MACLEOD, S., CASINADER, R. and MCGEE, T.G. (1987) "The Last Frontier. The Emergence of a Global Industrial Food System: A Case Study of Hong Kong." Unpublished paper.

MAGNUSSON, W. (ed.) (1984) *The New Reality: The Politics of Restraint in British Columbia.* Vancouver: New Star Books.

MAHER, C.A. and MCKAY, J. (1985) *Temporal trends in Australia's internal migration: the accumulating evidence, 1966-1981.* Department of Geography, Monash University, 1981 Internal Migration Study, Working Paper No. 10.

MANDEL, E. (1975) *Late Capitalism.* London: New Left Books.

MANSOUR, V. (1987) "From the bottom up" *Atlantic Insight* March, pp. 23-27.

MARCHAND, C. (1986), "The transmission of fluctuations in a central place system" *Canadian Geographer* 30, pp. 249-54.

MARSHALL, J.N. (1985) "Research policy and review. Services in a postindustrial economy" *Environment and Planning A* 17, pp. 1155-67.

MARTIN, R.L. (1984.) "Redundancies, labour turnover and employment contraction in the recession: a regional analysis" *Regional Studies* 18, pp. 445-458.

MASSEY, D. (1984) *Spatial Divisions of Labour: Social Structures and the Geography of Production.* London: Macmillan.

MASSEY, D.B. and ALLEN, J. (eds.) (1984) *Geography Matters!* Cambridge: Cambridge University Press.

MATHIAS, P. (1971) *Forced Growth.* Toronto: James Lewis and Samuel.

MATTHEWS, R. A. (1981) "Two alternative explanations of the problem of regional dependency in Canada" *Canadian Public Policy* 7, pp. 268-83.

MATTHEWS, R. A. (1983) *The Creation of Regional Dependency,* Toronto: University of Toronto Press.

MATTHEWS, R. A. (1985) *Structural Change and Industrial Policy: The Redeployment of Canadian Manufacturing, 1960-80.* Ottawa: Economic Council of Canada.

MAULDEN, F.R.E. (1929) *The Economics of Australian Coal* Melbourne: Melbourne University Press.

McCANN, L.(1981) "Heartland and hinterland: a framework for regional analysis", in McCANN, L.D. (ed) *A Geography of Canada: Heartland and Hinterland.* Scarborough: Prentice-Hall Canada, pp. 1-35.

McCARTY, J. (1970) "Australian capital cities in the nineteenth century" *Australian Economic History Review* 10, pp. 107-137.

MCGEE, T.G. (1983) *Canada and the Changing Economy of the Pacific Basin. An Introductory Overview.* Working Paper, No. 1 Institute of Asian Research, UBC: Vancouver.

MCGEE, T. (1985) "Women Workers or Working Women?" Unpublished paper, University of British Columbia, Department of Geography.

MCGEE, T.G. (1986) "Circuits and networks of capital: the internationalization of the world economy and national urbanization" in Drakakis-Smith, D. (ed.) *Urbanization in the Developing World.* London: Croom Helm, pp. 23-36.

MCGUINNESS, P.P, (1987) "Debt is the core of the problem', 'Cutbacks needed to speed adjustments', Protection racket has to be stopped" *Australian Financial Review* December, pp. 7-9.

MCKAY, J. (1984) "Australian migration and overview" in *Migration in Australia.* Brisbane: Royal Geographical Society (Qld.) and Australian Population Association, pp. 21-46.

MCKAY, J. (1985) "Internal migration and rural labour markets" in POWELL, R. (ed.) *Rural Labour Markets in Australia.* Canberra: AGPS, pp. 160-207.

MCKAY, J. and WHITELAW, J.S. (1977) "The role of large private and government organizations in generating flows of interregional migrants: the case of Australia" *Economic Geography* 53, pp. 28-44.

MCKAY, J. and WHITELAW, J.S. (1978) "Internal migration and the Australian urban system" *Progress in Planning* 10, pp. 1-83.

MCLOUGHLIN, P. (1985) "Industrial restructuring and its relationship to small business in non-metropolitan Victoria" *Journal of Australian Political Economy* 19, pp. 49-61.

MCLOUGHLIN, P. (1986a) "Trends and Rationale for Federal Regional Policy in Australia" paper presented to the *Joint Meeting of the Industrial Study Groups of Canada and Australia*, Vancouver, Canada, 28pp.

MCLOUGHLIN, P. (1986b) *Structural Change in Australian Regions: A Shift Share Analysis.* Country Centres Project Discussion Paper, Canberra: DOLGAS.

MCMICHAEL, P. (1984) *Settlers and the Agrarian Question: Foundations of Capitalism in Colonial Australia.* Cambridge: Cambridge University Press.

MCMILLAN, C.J. (1981) "The pros and cons of a national export trading house" *Canadian Public Policy* 7, pp. 569-83.

MCNIVEN, J. D. (1986) "Regional development policy in the next decade" *Canadian Journal of Regional Science* 9, pp. 78-88.

MCNIVEN, J. D.(1987) "The efficiency-equity tradeoff: the Macdonald Report and regional development", in COFFEY, W.J. and POLESE, M. (eds.) *Still Living Together: Recent Trends and Future Directions in Canadian Regional Development.* Halifax: Institute for Research on Public Policy, pp. 425-36.

MILLS, C. (1988) "Life on the upslope': the postmodern landscape of gentrification" *Society and Space* 6, pp. 169-90.

MILLS, C. (1989) *Interpreting Gentrification: Postindustrial, Postpatriachal, Postmodern?* Unpublished Ph.D. Thesis, University of British Columbia, Department of Geography.

MINISTRY OF EDUCATION (1988) *Technology Business Incubators and Tertiary Institutions.* Victoria: Triple I Technology Consultants and Norman Wheeler.

MITTER, S. (1986) "Industrial restructuring and manufacturing homework" *Capital and Class* 27, pp.37-80.

MORRIS, P.J. (1983) "Australia's dependence on imported technology - some issues for discussion" *Prometheus* 1, pp. 144-159.

NATIONAL POPULATION INQUIRY (1985) *Population and Australia.* Canberra: AGPS.

NAYLOR, R.T. (1975) *The History of Canadian Business 1867-1914*. Toronto: James Lorimer.

NELLES, H. V. (1974) *The Politics of Development: Forests, Mines & Hydro- electric Power in Ontario, 1849-1941.* Toronto: Macmillan of Canada.

NEW SOUTH WALES JOINT COAL BOARD (1947-1985) *Annual Reports*. Sydney: Government Printers.

NEW ZEALAND FOREST PRODUCTS LIMITED (1982) *Annual Report and Accounts.*

NEWMAN, P. (1982) *The Canadian Establishment Vol. 2. The Acquisitors.* Toronto: Seal Books.

NINOMI, S. (1985) "Japan has become a major supplier of capital to the world" *Japan Economic Almanac* pp. 23-24.

NIOSI, (1985) *Canadian Multinationals.* Toronto: Between the Lines.

NORCLIFFE, G.B. (1987) "Regional unemployment in Canada in the 1981-

1984 recession" *Canadian Geographer* 31, pp. 150-59.

NORCLIFFE, G.B. (1988) "Industrial structure and labour market adjustments in Canada during the 1981-84 recession," *Canadian Journal of Regional Science* 11, pp. 201-26.

NORRIE K. (ed.) (1986) *Disparities and Interregional Adjustment.* Research Study #64, Royal Commission on the Economic Union and Development Prospects for Canada, Toronto: University of Toronto Press.

NORRIE, K. and PERCY, M. (1983) "Freight rate reform and regional burden" *Canadian Journal of Economics* 16, pp. 325-349.

NORRIE, K.H. and PERCY, M.B. (1982) "Province-Building and Industrial Structure in a Small Open Economy." Paper prepared for *The Second John Deutsch Roundtable on Economic Policy.* Kingston, Ontario: Queen's University.

NOYELLE, T.J. and STANBECK, T.M. (1983) *The Economic Transformation of American Cities.* Ottawa: N.J. Rowman and Allanheld.

O'CONNOR, K. (1986) "The Restructuring Process Under Constraints: A Study of Recent Economic Change in Australia" paper presented to *The Regional Science Association, Australia and New Zealand Section*, Sydney, Australia, 26pp.

O'NEILL, P.M. (1986) *National Economic Change and The Labour Process in a Non-metroplitan Region: Dubbo, New South Wales*, Unpublished MA (Hons) Thesis, Macquarie University.

OECD (1985) *Reviews of National Science Policies: Australia.* Paris.

OECD, (1983) *Positive Adjustment Policies: Managing Structural Change.* Paris.

OECD, (1984) *Australian Urban Economic Development.* Group on Urban Affairs. Paris.

OLSEN, M. (1982) *The Rise and Decline of Nations.* New Haven: Yale University Press.

ONTARIO, (1967) *Report of the Ontario Committee on Taxation.* Toronto: Queen's Printer.

ORGANIZATION FOR ECONOMIC COOPERATION AND DEVELOPMENT (1985) *Review of Australian Science and Technology: Draft Reports.* Paris: OECD.

OVERTON, J. (forthcoming) "The crisis and the 'small is beautiful' model of development" in FAIRLEY, B., LEYS, C. and SACOURMAN, J.

(eds.) *Restructuring and Resistance in Atlantic Canada*, Toronto: University of Toronto Press.

PARIS, C. (1985) "Housing issues and policies in Australia" *Built Environment* 11, pp. 97-116.

PARRY, T.G. (ed) (1982) *Australia Industry Policy*, Melbourne: Longman Cheshire.

PATIENCE, A. and SCOTT, J. (eds) (1983) *Australian Federalism: Future Tense*. Melbourne: Oxford University Press.

PAVITT, K. and WALKER, W. (1976) "Government policies towards industrial innovation: A review" *Research Policy* 5.

PERCY, M.B. (1986) *Forest Management and Economic Growth*. Ottawa: Economic Council of Canada.

PERRONS, D.C. (1981) "The role of Ireland in the new international division of labour: a proposed framework for regional analysis" *Regional Studies* 15, pp. 81-200.

PIORE, M.J. and SABEL, C.F. (1984) *The Second Industrial Divide*. New York: Basic Books.

PITEMAN, H. (1982) *The Effectiveness of Victorian Decentralization Incentives on Enterprise Location Decisions*. Victoria Chamber of Manufacturers, Discussion Paper No. 9, VCM, Melbourne.

POLESE, M. (1981) "Regional disparity, migration and economic adjustment: a reappraisal" *Canadian Public Policy* 7, pp. 519-525.

POWDITCH, T. (1985) "Rise of the 'big daddies' " *Australian Business* 30 October, pp.114-7.

POWER, M. (1980) "Unemployed women: scapegoats of the recession" in Crough, G. (ed.) *Australia and World Capitalism*. Ringwood: Penguin.

PRATT, G. (1984) *An Appraisal of the Incorporation Thesis: Housing Tenure and Political Values in Urban Canada*, Unpublished Ph.D. Thesis, University of British Columbia, Department of Geography.

PRATT, G. (1986) "Housing-Consumption Sectors and Political Response in Urban Canada," *Environment and Planning Society and Space* 4, pp. 165-82.

QUEENSLAND COAL BOARD (1950-51 to 1984-85) *Annual Reports*. Brisbane: Government Printers.

QUEENSLAND DEPARTMENT OF MINES (1950-1985) *Annual Reports*. Brisbane: Government Printers.

RABAN, J. (1974) *Soft City*, New York: E.P. Dutton.

RABEAU, Y. (1987) "Regional efficiency and problems of financing the Canadian federation" in COFFEY, W.J. and POLESE, M. (eds.) *Still Living Together: Recent Trends and Future Directions in Canadian Regional Development.* Halifax: Institute for Research on Public Policy, pp. 383-406.

RAYNAULD, A. (1986) "Les politiques de croissance" *Canadian Public Policy* 12, Supplement, pp. 68-75.

RAYNAULD, J. (1987) "Canadian regional cycles and the propagation of US economic conditions" *Canadian Journal of Regional Science* 10, pp. 77-89.

REISMAN, S. (1985) "Canada-United States trade at the crossroads: options for growth" *The Canadian Business Review*, Autumn, pp. 17-23.

RESEARCH SERVICE (1950) *Report on Coal, 1950.* Sydney: Stewart Howard and Associates Pty. Ltd.

RICH, D.C. (1987) *The Industrial Geography of Australia*, North Ryde: Metheun.

RICH, D.C. (1988) "Australian manufacturing, 1968-1988: recession, reorganisation and renewal," *GeoJournal* 16, 399-418.

RICHARDS, J. and PRATT, L. (1979) *Prairie Capitalism: Power and Influence in the New West.* Toronto: McClelland and Stewart.

RIMMER, P. (1986) "Australian and Japanese international engineering consultancy services" *Pacific Economic Papers, No 131.* Canberra: Australia - Japan Research Centre.

RIMMER, P.J. (1986) "Japan's world cities: Tokyo, Osaka, Nagoya or Tokaido Metropolis?" *Development and Change* 17, pp.121-159.

ROBINS, B. (1986) "OTC announces $2b modernization plan" *Australian Financial Review* 27 June, p.33.

ROCHE, M.M. (1987a) "Company afforestation: patterns and processes during the 'first planting boom' " *Proceedings 13th New Zealand Geography Conference.*

ROCHE, M.M. (1987b) "New Zealand Timber for the New Zealanders. Regulatory Controls and the Decline of the New Zealand-Pacific Rim Timber Trade in the 1920s and 1930s" *Proceedings 14th New Zealand Geography Conference.*

RONAYNE, J. (1984) *Science in Government.* Caulfield East, Australia: Edward Arnold.

ROSS, D.A. (1982) "Canadian foreign policy and the Pacific rim: from national security anxiety to creative economic cooperation" in QUO, F.Q. (ed.), *Politics of the Pacific Rim: Perspectives on the 1980s*. Burnaby, Simon Fraser University, pp. 27-58.

ROTHWELL, R. (1983) *Design and the Economy*. London: The Design Council.

ROTHWELL, R. (1986) "Innovation and re-innovation: a role for the user" *Journal of Marketing Management* 2, pp. 109-123.

ROTHWELL, R. and ZEGFELD, W. (1985) *Reindustrialization and Technology*. London: Longman.

RURAL DEVELOPMENT CENTRE (1985) "On the road" *Inside Australia* 1, pp. 8-10.

RUTLEDGE, I. and WRIGHT, P. (1985) "Coal worldwide: the international context of the British miners' strike" *Cambridge Journal of Economics* 9, pp.303-326.

SAFARIAN, A.E. (1985) *FIRA and PIRB: Canadian and Australian Policies on Direct Foreign Investment*. Toronto: Ontario Economic Council.

SANT, M. (1978) "Issues in employment" in DAVIES, R. and HALL, P. (eds.) *Issues in Urban Society*. Hardmondsworth, Middlesex: Penguin, pp. 84-105.

SARGENT, S. (1985) *The Foodmakers*. Ringwood, Vic: Penguin.

SASSEN KOOB, S. (1984) "The new labour demand in global cities" in SMITH M.P. (ed.) *Cities in Transformation Class, Capital and the State* Vol. 26, Urban Affairs Annual Reviews. Beverly Hills, Sage Publications, pp.139-171.

SAVOIE, D. (1981) *Federal-Provincial Collaboration: the Canada - New Brunswick General Development Agreement*. Montreal: McGill - Queen's University Press.

SAVOIE, D. (1986a) *Regional Economic Development: Canada's Search for Solutions*. Toronto: University of Toronto Press.

SAVOIE, D. J. (1984) "The toppling of DREE and the prospects for regional economic development" *Canadian Public Policy* 10, (3), pp. 328-337.

SAVOIE, D.J. (1986b) "Courchene and regional development: beyond the neoclassical approach" *Canadian Journal of Regional Science* 9, pp. 69-78.

SAYER, A. (1985) "Industry and space: A sympathetic critique of radical research" *Environment and Planning D: Society and Space* 3, pp. 3-29.

SAYER, A. (1986) "Industrial location on a world scale: the case of the semiconductor industry" in SCOTT, A. J. and STORPER, M. (eds.) *Production, Work, Territory.* London: Allen and Unwin, pp. 107-123.

SAYER, A. and MORGAN, K. (1985) "A modern industry in a declining region: links between method, theory and policy" in MASSEY, D. and MEEGAN, R. (eds.) *Politics and Method: Contrasting Studies in Industrial Geography.* London: Methuen, pp.144-168.

SCHOENBERGER, (1988) "From Fordism to flexible accumulation: technology, competitive strategies and international location: *Society and Space* 6, pp. 245-262.

SCIENCE COUNCIL OF CANADA (1980) *Multinationals and Industrial Strategy: The Role of World Product Mandates.* Ottawa.

SCOTT, A. D. (ed.) (1975) *Natural Resource Revenues: a Test of Federalism, Vancouver*: University of British Columbia Press

SCOTT, A. J. (1986) "Industrialization and urbanization: a geographical agenda" *Annals of the Association of American Geographers* 76, pp.25-37.

SCOTT, A.J. (1983) "Industrial organization and the logic of intra-metropolitan location: I. Theoretical considerations" *Economic Geography* 59, pp. 233-25O.

SCOTT, A.J. and STORPER, M. (eds.) (1986) *Production, Work, Territory. The Geographical Anatomy of Industrial Capitalism.* Boston: Allen and Unwin.

.SCUTT, J. (1985) (ed.) *Poor Nation of the Pacific: Australia's Future.* Sydney: Allen and Unwin.

SEARLE, G.H. (1985) High Technology Industry Location and Planning Policy in the Sydney Region, unpublished paper, Innovation, Technological Change and Spatial Impacts Workshop, CSIRO, August 1985.

SEGAL, QUINCE, WICKSTEED (1985) *The Cambridge Phenomenon.* London: In Association with Brand Brothers and Co.

SHEARER, R. A. (1986) "The new face of Canadian mercantilism: the Macdonald Commission and the case for free trade" *Canadian Public Policy* 12, Supplement, pp. 51-58.

SHUMPIE, K. (1986) "Designs for a new industrial society" *Japan Echo* 13, Special Issue, pp. 1-3.

SIMEON, R. (1987) "Inside the Macdonald Commission" *Studies in Political Economy* 22, Spring, pp. 167-79.

SLOAN, J. and KRIEGLER, R. (1984) "Technological change and migrant employment" *Australian Quarterly* 56, pp. 216-26.

SMITH, C.A. (1985) "Theories and measures of urban primacy: a critique" in TIMBERLAKE, M. (ed.) *Urbanization in the World-Economy.* Academic Press.

SMITH, D. A. (1984) "The development of employment and training programs" in MASLOVE, A.M. (ed.) *How Ottawa Spends: The New Agenda,* Toronto: Methuen 1984, pp.167-88.

SNUKAL, S. (1982) *Talking Dirty,* Toronto: Playwrites Canada.

SOUTH AUSTRALIAN GOVERNMENT, (1984) *Some Comments on Medium and Longer-Term Economic Growth Strategies.* Canberra: Economic Planning Advisory Council.

STANBACK, T. and NOYELLE, T. (1982) *Cities in Transition,* Ottawa, N.J.: Allenheld, Osmun and Co..

STATISTICS CANADA (1978) *Canadian Imports by Domestic and Foreign Controlled Enterprises.* Ottawa: Cat. No. 67-509.

STEED, G.P.F., and deGENOVA, D. (1983) "Ottawa's technology-oriented complex" *Canadian Geographer* 27, pp.263-278.

STEKETEE, M. (1986) "Button hangs on for the long haul" *Sydney Morning Herald* Friday, May 30, p. 26.

STILWELL, F.J.B. (198O) *Economic Crisis, Cities and Regions.* Pergamon Press: Sydney.

STILWELL, F.J.B. (1983a) "Is there an Australian economy?" in ALDRED J. AND WILKES, J. (eds) *A Fractured Federation?*, Sydney: Allen and Unwin, pp. 19-36.

STILWELL, F.J.B. (1983B) "State and capital in urban and regional development" in HEAD, B.W. (ed) *State and Economy in Australia.* Melbourne: Oxford University Press.

STILWELL, F.J.B. (1985) "The Integration of Regional Policy and Industry Policy" paper presented to *the National Conference on Australian Industry and the Future,* Wollongong, Australia.

STORER, D. (198O) "Migrants and unemployment" In CROUGH, G., WHEELWRIGHT, T., and WILSHIRE, T. (eds.) *Australia and World Capitalism.* Ringwood: Penguin, pp. 46-52.

SUSMAN, P. and SCHUTZ, E. (1983) "Monopoly and competitive firm relations and regional development in global capitalism," *Economic Geography,* 59, pp. 161-177.

SWEENEY, G.P. (1987) *Innovation, Entrepreneurs and Regional Development.* London: Frances Pinter.

TARASOFSKY, A. (1984) *The Subsidization of Innovation Projects by the Government of Canada.* Ottawa: Economic Council of Canada.

TAYLOR, G. (1951) *Australia: A Study of Warm Environments and their Effect on British Settlement.* New York: Methuen, sixth enlarged edition.

TAYLOR, J. and BRADLEY, S. (1983) "Spatial variations in the unemployment rate: a case study of North West England" *Regional Studies* 17, pp. 113-124.

TAYLOR, M. and THRIFT, N.J. (1981) "Some geographical implications of foreign investment in the semi-periphery: the case of Australia" *Tijdschrift voor Economische en Sociale Geografie* 72, pp. 194-213.

TAYLOR, M. and THRIFT, N.J. (1983) "Business organization, segmentation and location" *Regional Studies* 6, pp. 445-465.

TAYLOR, M.J. and THRIFT, N. (198O) "Large corporations and concentrations of capital in Australia" *Economic Geography* 55, pp.261-180.

THOMAS, P. (1983) *Miners in the 1970s: A Narrative History of the Miners Federation* Sydney: Miners Federation.

THRIFT, N. (1986) "The internationalization of producer services and the integration of the Pacific Basin property market" in Taylor M. and Thrift, N. (eds.) *Multinationals and the Restructuring of the World Economy*, London: Croom Helm.

THRIFT, N.J. (1986) "The geography of international economic disorder", in JOHNSTON, R.J. and TAYLOR, P.J. (eds.) *A World in Crisis?* Oxford: Blackwell, pp. 12-67.

TIMBERLAKE, M. (1985) "The world-system perspective and urbanization" in TIMBERLAKE, M. (ed.) *Urbanization in the World-Economy.* Academic Press.

TNC WORKERS' RESEARCH (1985) *Anti-union Employment Practices: Final Report.* Sydney: TNC Workers' Research.

TOFFLER, A. (1985) *The Adaptive Corporation.* London: Pan Book.

TOMLINSON, J.W.C. and HUNG, C.L. (1983a) "A profile of Canadian corporate activities in the Asian Pacific countries" in *Canada and Changing Economy of the Pacific Basin,* Working Paper No. 2, Institute of Asian Research, University of British Columbia.

TOMLINSON, J.W.C. and HUNG, C.L. (1983b) "The Investment and operational characteristics of Canadian Companies in the Asian Pacific Region", *Canada and the Changing Economy of the Pacific Basin*, Working Paper No. 9, Institute of Asian Research, University of British Columbia.

TOWNSEND, A.R. (1983) *The Impact of Recession on Industry and Employment, and the Regions 1976-1981*, London: Croom Helm.

TRACHTE, K. and ROSS, R. (1985) "The crisis of Detroit and the emergence of global capitalism" *International Journal of Urban and Regional Research* 9, pp. 186-216.

TUPPER, A. and DOERN G. B. (eds.) (1981) *Public Corporations and Public Policy in Canada.* Montreal: Institute for Research on Public Policy.

US (1980) *International Data Flow*, Hearings before a Subcommittee of the Committee on Government Operations, Washington, D.C.: House of Representatives, 96th Congress, second session, March and April.

US (1984) *US National Study on Trade in Services*, Washington, D.C.: A submission by the United States Government to the General Agreement on Tariffs and Trade.

US INTERNATIONAL TRADE ADMINISTRATION (1984) *A Competitive Assessment of the US Data Processing Services Industry*, Washington D.C.: US Department of Commerce.

US NATIONAL TELECOMMUNICATIONS AND INFORMATION AMINISTRATION (1983) *Long-Range Goals in International Telecommunications and Information: An Outline for United States Policy*, Washington, D.C. Report to the Congress of the US, February.

VANDERKAMP, J. (1986) "The efficiency of the interregional adjustment process," in NORRIE, K. (research co-ordinator), *Disparities and Interregional Adjustment.* Research Study #64, Royal Commission on the Economic Union and Development Prospects for Canada. Toronto: University of Toronto Press, pp. 53-108.

VEAC (VANCOUVER ECONOMIC ADVISORY COMMISSION) (1984) *Changing Skills: A Survey on Vancouver's Skill Requirements in the Eighties*, Vancouver.

VERNON, R. (1977) *Storm Over The Multinationals.* Cambridge: Harvard University Press.

WADLEY, D. (1986) *Restructuring the Regions: Analysis, Policy Model and Prognosis*, Paris: OECD.

WALKER, D. (198O) "Political aspects of regional industrial development" in WALKER, D.F. (ed) *Planning Industrial Development*. Winchester: Wiley, pp. 129-148.

WALMSLEY, D.J. (198O) "Welfare delivery in post-industrial society" *Geografiska Annaler* 62B, pp. 91-7.

WALMSLEY, D.J. (1984) "Fiscal equalization and Australian federalism 1971-81" *Environment and Planning C" Government and Policy* 2, pp. 93-106.

WALTON, J. (ed.) (1984) *Capital and Labour in the Urbanized World.* London: Sage.

WATERS, N. (1985) *Structural Change in the British Columbia Economy*, Unpublished M.B.A., Thesis, University of British Columbia, Faculty of Commerce.

WATKINS REPORT (1968) *Foreign Ownership and the Structure of Canadian Industry*. Ottawa: Privy Council Office.

WATKINS, M.H. (1963) "A staple theory of economic growth" *The Canadian Journal of Economic and Political Science* 29, pp. 141-58.

WATKINS, M.H. (1973) "Resources and underdevelopment" in LAXER, R. (ed.) *Canada Ltd. The Political Economy of Dependency.* Toronto: McClelland and Stewart, pp.107-126.

WATKINS, M.H. (1977) "The staple theory revisited" *Journal of Canadian Studies* 12, pp.83-95.

WATKINS, M.H. (1978) "The economics of nationalism and the nationality of economics: a critique of neoclassical theorizing" *Canadian Journal of Economics* 11, pp. S87-S119.

WATSON, W.G. (1987) "Canada-US free trade: why now?" *Canadian Public Policy* 13, pp. 337-49.

WEAVER, C. (1985) "Regions, decentralization and the new global economy: an overview" *Canadian Journal of Regional Science* 8, pp. 283-97.

WEBBER, M.J. (1982) "Agglomeration and the regional question" *Antipode* 14, pp.1-11.

WELCH, R. (1982) "Spatial requirements for the continuation of social democracy in New Zealand" *New Zealand Geographer* 38, pp. 3-8.

WEST, K. (1984) *The Revolution in Australian Politics*, Ringwood: Penguin.

WHALLEY, J. AND TRELA, I. (1986) *Regional Aspects of Confederation*, Research Study #68, Royal Commission on the Economic

Union and Development Prospects for Canada, Toronto: University of Toronto Press.

WHEELWRIGHT, T. (198O) "The age of the transnational corporation". In CROUGH, G., WHEELWRIGHT, T. and WILSHIRE, T. (eds.) *Australia and World Capitalism.* Ringwood: Penguin, pp. 123-36.

WILDE, P.D. (1981) "From insulation towards integration: the Australian industrial system in the throes of change" *Pacific Viewpoint* 22, pp.1-24.

WILDE, P.D. (1986) "Economic restructuring and Australia's changing role in the world economic system" in HAMILTON, F.E.I. (ed.) *Industrialization in Peripheral and Developing Countries.* London: Croom Helm, pp.16-43.

WILDE, P.D. and FAGAN, R.H. (1988) "Industrial geography: restructuring in theory and practice," *Australian Geographical Studies* 26, 132-148.

WILSON, R.K. (1978), "Urban and Regional Policy" in SCOTTON, R.B. and FURBER, H. (eds) *Public Expenditures and Social Policy in Australia: Volume l, The Whitlam Years 1972-75*, Melbourne: Longman Cheshire, pp. 179-211.

WINDSCHUTTLE, K. (1984) "High tech and jobs" *Australian Society* 3, pp. 11-3.

WITHERS, T. (1986) "Rada denies pulp rationalization plans" *Dominion*, 10 April, p.12.

WOCOL, (1980) *Coal-Bridge to the Future: Report of the World Coal Study.* Cambridge, Massachussetts: Ballinger Publishing Company.

WOLF, E.R. (1982) *Europe and the People Without History.* Berkeley: University of California Press.

WONOROFF, J. (1985) *Japan's Wasted Workers.* Tokyo: Lotus Press.

YUTAKA, K. (1986) "Patterns of change in Japan's industrial society" *Japan Echo* 13, Special Issue, pp. 32-39.